ALLE ZEIT WACH
1842

Charles Devillers
Jean Chaline

Evolution

An Evolving Theory

With 76 Figures

Springer-Verlag
Berlin Heidelberg New York
London Paris Tokyo
Hong Kong Barcelona
Budapest

Dr. Charles Devillers
25, rue de Bagneux
F-92330 Sceaux
France

Professor Dr. Jean Chaline
Université de Bourgogne
Centre des Sciences de la Terre
6, Bd. Gabriel
F-21000 Dijon
France

Translated by
Dr. Thomas Reimer
Via Lavizari 2a/BSM
CH-6900 Lugano
Switzerland

Title of the original French edition:
Charles Devillers et Jean Chaline: La théorie de l'évolution

ISBN 3-540-54674-X Springer-Verlag Berlin Heidelberg New York
ISBN 0-387-54674-X Springer-Verlag New York Berlin Heidelberg

Library of Congress Cataloging-in-Publication Data. Devillers, Charles, 1914– Evolution: an evolving theory / Charles Devillers, Jean Chaline. p. cm. Includes bibliographical references (p.) and index.
ISBN 3-540-54674-X (Berlin). –
ISBN 0-387-54674-X (New York)
1. Evolution (Biology) I. Chaline, Jean. II. Title. QH366.2.D346 1993 575–dc20 93-8408 CIP

Printed in Germany

Cover Design: Struve&Partner, Atelier für Grafik-Design, Heidelberg
Typesetting: Data conversion by Springer-Verlag
32/3145-5 4 3 2 1 0 – Printed on acid-free paper

We dedicate this book
to those who have contributed to the elaboration
of a new, more comprehensive and explicative theory
of the evolution of life

Preface

Biological evolution is a discipline about which scientists and laymen alike feel authorized to make more or less learned statements. However, the study of evolution and its processes and patterns requires the understanding of such diverse disciplines of natural sciences as genetics, embryology, ecology, palaeontology, etc. We have written this book with the aim of presenting some indispensible basic data on the subject to an open-minded, but frequently insufficiently informed public. By its very design, the book cannot treat the subject exhaustively and we have had to omit a number of aspects. We could be reproached for this, and we accept the responsibility for these deficits, of which we certainly are aware.

The synthetic theory of evolution was elaborated in the Anglo-Saxon countries over the past 40 years from a combination of genetic, biological, and palaeontological data. With the exception of Philippe L'Heritier and George Tessier, the synthetic theory was rejected by a number of French specialists who curiously did not accept chance and natural selection as the main driving forces of evolution. They rather adhered either to a neo-Lamarckism or to a vitalistic trend which received some support from authorities like Henri Bergson and, especially, Teilhard de Chardin, or to a concept of evolution favouring major changes in the structure of organisms. This latter concept is referred to as "macroevolution" following the work of Richard Goldschmidt, Albert Vandel, and Pierre-Paul Grassé.

Since then genetics, biology, and palaeontology have achieved tremendous progress and have cast doubts on a number of aspects of the synthetic theory, which by some authors has been considered rather prematurely as a dogma. The discoveries of molecular biology, the establishment of the genetics of development, new concepts of speciation, and the analysis of spatial and temporal relationships of evolution in palaeontology are major steps in this progress. Furthermore, we would like to stress the stimulating role currently played by the palaeontologist Stephen J. Gould. His views, to which we do not necessarily subscribe completely, have initiated an international debate on the re-evaluation of these concepts.

On the threshold of the coming century, the innovative aspects of the theory of evolution have become so numerous that the time appears ripe for a new assessment of the concept. Such a new synthesis requires the input from more than one specialist. The fields implicated are so varied and the wealth of data furnished by the sciences of life, Earth, and the Universe are so overwhelming that it becomes increasingly difficult to know and master them all. As a consequence, a biologist and a palaeontologist have combined forces to produce this book. The treatment is far from exhaustive and is directed at a public interested in the following major questions: What is the place of man in the Universe and in the living world? What is evolution? How can the observed data be explained? Where do we stand with the theory of evolution?

With this in mind, we have intentionally left out certain aspects so that we may answer those questions which are put to us by a public with different degress of knowledge. This approach will permit us to outline the major trends of a new state of the evolutionary synthesis in a critical approach supported by examples.

This book relies heavily on the results of the research group "Analytical Palaeontology" at the CNRS No. 157 at Dijon in France. This laboratory was founded in the 1960s by Henri Tintant who introduced to French palaeontology the concept of populations already used in studies of living species. His statistical methods are still much in use. The group was subsequently expanded and enlarged its field of investigation. It devoted itself to the in-depth studies of various aspects of the palaeontological approach to evolution, with the aim of eventually discussing the evolutionary concepts of the new synthetic theory. This book also owes much to the weekly seminars held at the laboratory, during which the most recent results of palaeontology and biology were critically evaluated. The book is furthermore a show-case for the Dijon Group, which contributed to its origin by critically reading the manuscript and suggesting improvements wherever they were considered necessary.

We want to extend our particular thanks to the following coworkers in Dijon: Bruno David, Jean-Louis Dommergues, Didier Marchand, Bernard Laurin, Denis-Didier Rousseau, and Henri Tintant. We are also grateful to Christine Rousseau and Christiane Fourcault who typed part of the manuscript, and to Annie Bussiere who prepared some of the drawings.

Furthermore, we have received invaluable help from colleagues in other specialty fields and from other laboratories. Amongst these we would like to name especially Michel Delsol, director of the EPHE laboratory of Post-Embryonal Development of Lower Vertebrates in Lyon for his comments on the role of chance, and Jean-Paul Parisot, professor of astronomy at the Besancon and Bordeaux observatories for his corrections to our historical framework of evolution.

Dr. Pierre Dusserre, director of the Laboratory of Human and Experimental Pathology in Dijon, together with Francoise Demoulin, and Renée and Jean-Marc Ridet, critically reviewed the text and provided valuable suggestions for improvements which help the understanding by the general public. Their assistance is gratefully acknowledged here.

And last, but not least, we would like to point out that the comments given by Laurent du Mesnil du Buisson, Marc Jammet, and Christine Manceron on the layout of this work greatly assisted us in preparing a book which would receive wide acceptance. The biological part of the book (Chaps. 3–6) was essentially written by C. Devillers, whereas J. Chaline prepared the palaeontological section (Chapts. 7, 9–14). Chapters 8 and 15 were written jointly and each chapter was critically read by the other co-author. To make reading easier, the bibliography, the appendices, and those examples which we consider indispensable for understanding our concepts are presented at the end of the book. They are referred to in the respective parts of the text in parentheses (e.g. Appendix 1).

Thus, the book is divided structurally into two levels. The first level, aimed at the greater public, is comprised of the first six chapters (Parts I and II) and the epilogue. The second level, represented by the appendices, is intended for those desiring more information and for specialists. A glossary will enable non-specialists to familiarize themselves with the scientific terms used.

The book is aimed at a public receptive to problems of the origin and evolution of life and especially of man himself, and also at teachers in various types of schools ("Towards a new evolutionary synthesis" is already one of the major new subjects at high schools). It should also interest biologists, palaeontologists, and specialists in various fields in the sciences of life, Earth, and the Universe, not to forget those who maintain a philosophical interest in the history of science.

Sceaux and Dijon, July 1993

C. Devillers
J. Chaline

Contents

Part III: The Message of Fossils from the Palaeontological Record

The Theory of Evolution – The 19th Century Concepts

The giraffe and its neck, as a symbol of evolution

Jean-Baptiste de Lamarck [1], who introduced the concept of transformism, interpreted the lengthening of the giraffe neck by the animal's desire to reach the higher leaves in trees: "As far as habits are concerned, it is of interest to observe its result, the particular shape of the giraffe *(G. camelopardalis)* and its size. We know that this tallest of all mammals lives in the African interior in areas which are almost continuously dry and without a grass cover. The animal thus has to live off the foliage of trees and has to strain itself continously to reach this foliage. As a result of this sustained habit, eventually in all individuals of this race, the front legs become longer than the hind legs, and the neck becomes lengthened to such an extent that the giraffe can lift its head to a height of six metres (close to 20 feet) without raising itself on its hind legs."

Desire and utilization are thus the moving forces of the transformation of the species and of evolution (Fig. 1).

With Charles Darwin [2], who introduced the concept of natural selection, another scenario was set. The various populations of giraffes possess a certain variability with short-, medium-, and long-necked individuals. As longer necks present several advantages for the individuals concerned, their chances of survival improve with each generation and they will have more descendants. As a consequence, natural selection will gradually shift the population to more individuals with still longer necks: "The giraffe, by its lofty nature, much elongated neck, fore legs, head and tongue, has its whole frame beautifully adapted for browsing on the higher branches of trees. It can thus obtain food beyond the reach of the other Ungulata or hoofed animals inhabiting the same country; and this must be a great advantage to it during dearths . . . So under nature with the nascent giraffe, the individuals which were the highest browsers and were able during dearths to reach even an inch or two above the others will often have been preserved; for they will have roamed over the whole country in search of food." (Fig. 1 [3]).

How do we interpret this phenomenon now? This is the aim of this book (see also Epilogue).

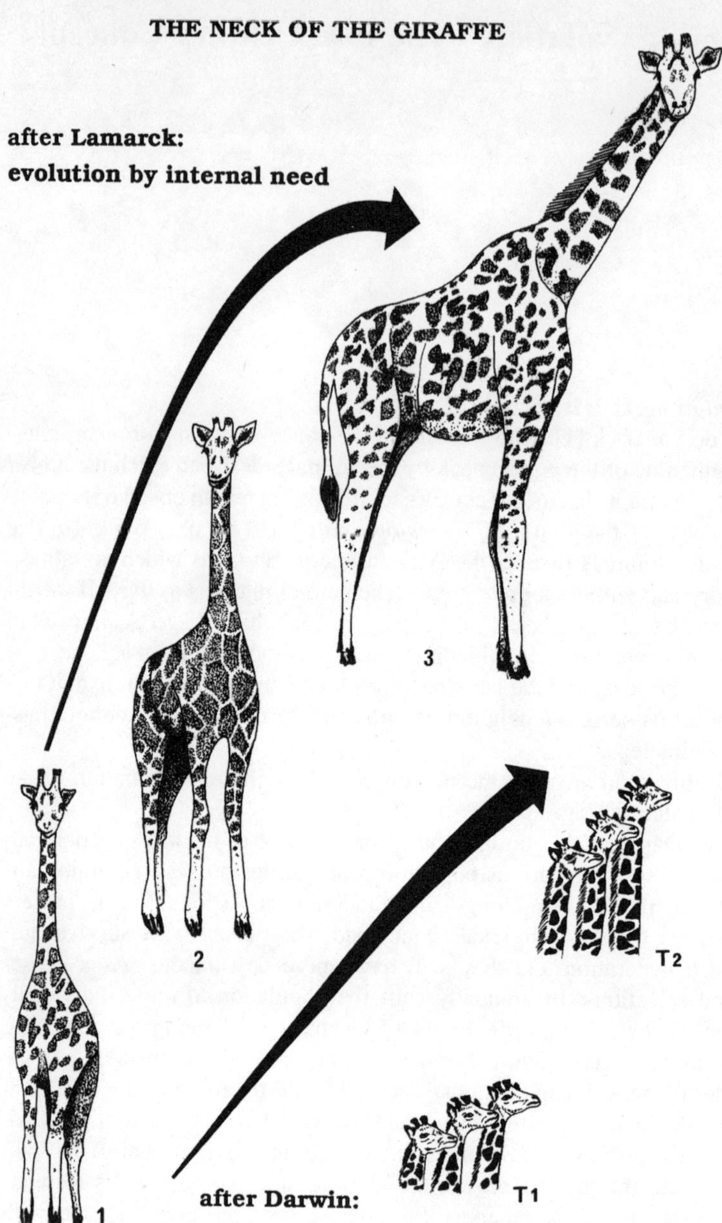

Fig.1. The neck of the giraffe as seen by Lamarck and Darwin. *1* According to Lamarck, the neck of the giraffe became longer as the result of an internal need of the animal to reach higher leaves in the trees. *2* According to Darwin, populations of giraffes possess a certain variability in neck length. Natural selection tended to favour animals with longer necks which could reach higher leaves in the trees and thereby secure a better chance of survival and reproduction. The selected feature was passed on to the following generations and, with time, the group exhibited a general increase in neck length. (After Savage [3])

Part I
Biological Evolution

Chapter 1

What Is Biological Evolution?

Development of the Concept of Evolution

Observers of the past and present noted an extraordinary diversity in the organic world: A snail is different from a river crayfish, which differs from a starfish, which in turn has nothing in common with a lobster or even man.

But let us take a closer, more detailed look: the snail, slug, and marine *Littorina* have certain common features of construction just as the river crayfish, lobster, and crab do.

Starting from the animal envisaged in its group, we observe increasingly finer levels of structure, namely the organs, tissues, and cells. We discover that all forms of life are made up of cells for which we are now able to outline a general plan of organization (Fig. 1.1).

Let us go further: to the molecular level, or to compounds like enzymes. We here find enzymatic mechanisms responsible for metabolic cycles which the simplest and most complex organisms have in common. And to complete the approach, molecular biology has revealed that all animals and plants posses a single, common generic code.

This leads us to an important conclusion: To understand the biological organization, we have to explain why theses similarities are there.

It will not suffice only to collect data. One also has to try to understand the information and the only valid scientific theory explaining the situation is the concept of evolution. It implies that there was an ancestral form far removed in age, already posessing some of the attributes which became features common to the entire organic world, such as genetic material, genetic code, cell organization, etc. From this, all types of organisms, from the simplest to the most complex, developed in succession, one deriving from the other. This concept of historic ties is the focal point of the theory of evolution. Could the history of the organic world, as revealed by palaeontology, assist us in finding an answer to this problem?

The oldest known organisms are bacteria (Fig. 1.2) from the 3.5- to 3.6-billion-year-old Archaean sediments of Western Australia. In the world as we know it now, bacteria are considered the simplest of all organisms, originating from the assemblage of "primitive molecules of biological interest." However, the degree of such an assemblage is much higher than that of any of these "primitive molecules."

Unicellular, more complex organisms [1], containing chlorophyll, appeared later, followed by multicellar organisms like jellyfish, arthropods, molluscs, and eventually vertebrates and man (Fig. 1.3). We thus have to acknowledge that there is an increase in organizational complexity with time, although more complex organisms do not always replace the less complex ones.

The concept of evolution is supported by studies in the past, revealing innumerable and formidable changes in organization. However, when looking at the world around us, we tend to accept a certain unchangeability: around us nothing appears to change, as we do not have the necessary yardstick at our disposal to measure duration, i.e. geological time.

Allowing for historical changes, we have to explain the respective evolutionary mechanisms. As we shall see later, our ideas regarding the mechanisms of these changes have themselves evolved since the beginning of the last century, and at present we are in a period of profound reassessment. The theory of evolution is by no means a dogma that claims to "explain the entire visible complexity by an invisible simplicity" (Jean Perrin). But does it still possess the character of a scientific theory, a character that it is denied by certain authors?

Defining Biological Evolution

Biological evolution is a process of hereditary, irreversible changes in the organic world. The respective transformations take place in certain time sequences in which one species follows another, becomes diversified, flourishes, and eventually becomes extinct.

The global advance of evolution is an increase in the structural and functional complexity of organisms, from the first ones to develop, the bacteria, to the most complex, such as molluscs, insects, and vertebrates (Figs. 1.3 and 1.4). This process of increasing complexity takes place neither regularly nor in one direction only. Increased complexity is acompanied by an increasingly diversified utilization of the environment.

The utilization of our planet's limited resources leads to competition between organisms and, thereby, to dominations (elimination of species) and also to the redistribution of habitats and resources between the various organisms involved. However, older forms, which are less complex than the younger ones, are not always eliminated and various patterns of coexistence develop. Although environmental conditions in part control organic transformations, thc rcvcrsc is also truc, for example in the formation of new ecological niches [2].

We have to envisage biological evolution under three complementary aspects: the mechanisms of organic transformation, the patterns of transformation with time, and their results through time. The analysis of the various mechanisms involved is the domain of present-day biologists (neontologists) as these mechanisms focus on genetic material, embryonic development, and populations, which can be studied only in living organisms. The patterns and the results, however, are observed especially over geological time spans. Present-day biologists do not have these time spans at their disposal, and thus these factors are studied by palaeontologists as the biologists of the past.

Biological evolution is not a process that goes in one direction. As George G. Simpson remarked, "evolution is opportunistic". As we shall show, evolution is a succession of events and thus is not predictable.

Is the Theory of Evolution a Scientific Theory?

Laymen frequently consider science as a world in which complete and definitive solutions are given to certain problems. All scientific disciplines influence each other and are directly dependent on the social and cultural context of their time.

In the advance of science, we may distinguish three phases. The first one concerns the clear definition of a problem, and the elaboration of methods and techniques to solve it. This is followed by the experimental phase, the gathering of observtions. The third phase involves the general, abstracted presentation of the results to form a concept, i.e. to model the phenomenon and to elaborate a theory.

A theory is a provisional, intellectual concept which should explain and also predict all known phenomena and observations concerning the problem in question. The discovery of a single new fact, not explained by the theory, will invalide it in part or completely. The theory will have to be replaced by a new one which takes the new data into account. A frequently problem of theoreticians is that they

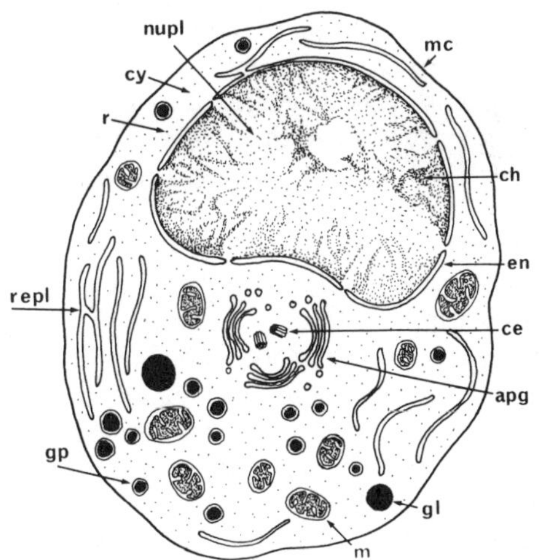

Fig.1.1. The eukaryotic cell. Simplified representation of a section under the electron microscope. Abbreviations: *apg* Golgi apparatus; *ce* centrioles; *ch* cromatin; *cy* cytoplasm; *en* porous nuclear membrane; *gl* and *gp* inclusions in cytoplasm; *m* mitochondria; *mc* cell membrane (actually double-walled); *nupl* content of nucleus or nucleoplasm; *r* ribosome; *repl* endoplasmic reticulum (occupying most of the cell volume)

try too rapidly to generalize a certain observation and to extend a general character to particular , or only partly valid, conditions.

The philosopher Karl Popper [3, 4], a protagonist of the critical study of theories in physics, biology, philosophy, and social sciences, concluded that the difference between metaphysical and scientific theories rests in the fact that the latter can be falsified. "The most far-reaching theory is the theory which can be most severely tested".

And: "We may conclude that scientific progress is not just an accumulation of observations, but the result of the rejection of less satisfying theories and their replacement by better ones, and in particular by more extensive theories".

For most authors, and especially for those from English-speaking countries, evolution is a fact. For us, however, the concept of evolution, of transformation is a theory, a mental concept which rearranges the observed facts in order to establish a causal connection between them. The content of the theory of evolution is therefore of a much higher explanatory value than the supposed factual nature of evolution.

Based on a rather simplistic, outdated view of the synthetic theory of evolution, K. Popper [3], in his original paper of 1978, considered the theory of evolution as a metaphysical one. Being more familiar with physical than with biological theories, he was strongly influenced by the fact that "Darwinism does not really predict evolution of the variability of species. It therefore is not able to really explain it." While crediting the synthetic theory only with the gradual and linear character of evolution, K. Popper recognizes a "prediction of the progressiveness" and the "accidental mutations". However, in doing so, he forgets one of the essential aspects of evolution, namely that it is subject to random events. Evolution is not predictable and as such contingent. As we shall show in Chapter 12, biological evolution does not respond to strict determinism. It is much more controlled by statistical indetermination like the quantum theory of atomic particles.

Wether it be scientific or metaphysical, we are tempted to ask "So what?" and gladly subscribe to the position of Pierre Thuillier [5] "A scientific theory is not a theory the truth of which has been definitively established by rational and intangible proof, but a stage in the process of organized systematic research." For the biologist of the past or present, the theory of evolution is a strong tool which is indispensable for investigating and understanding the organic world. Theodosius Dobzhansky [6] expressed this most aptly:" In biology nothing makes sense unless considered in the light of evolution."

The general theory of evolution actually consists of a whole group of theories which explain various aspects of the phenomenon and may be attacked at any analytical level. The theory of evolution, the synthesis of which we intend to present in a wider outline in this book, is thus a scientific theory in the sense of K. Popper. One could not say this about "creationism", which postulates the independent creation of the different species by God. This is a metaphysical theory which dogmatically refuses to be tested or verified, by rejecting any scientific data which might be in conflict with it.

In his analysis of the major scientific revolutions, especially in the physical sciences, Olivier Costa de Beauregard [7] has shown that scientific advances

PROKARYOTES Multiplication by fission	EUKARYOTES Sexual reproduction			
Unicellular		Multicellular Division of labor in cells		
MONERA	PROTISTS	METAPHYTES	FUNGI	METAZOANS
Bacteria Cyanobacteria	Protozoans Algae and unicellular fungi	"Higher" plants		"Higher" animals
Autotrophs Heterotrophs Osmiotrophs	Autotrophs Heterotrophs	Autotrophs	Saprophytes	Heterotrophs
	Mobility in Protozoans	No mobility		Mobility

Fig.1.2. New classification of organisms. In *prokaryotes,* multiplication takes place through binary fission. In bacteria, the phenomenin of conjugation represents a primitive „sexualitiy" whereby the chromosomes of one bacterium (here called „male") are injected into a „female" bacterium. Cell multiplication in *eukaryotes* is more complex, occuring through mitosis which entails the following steps: disappearance of the nucleus wall, doubling of the number of chromosomes, reconstitution of two nuclei with a full complement of chromosomes, and division into two daughter cells. Sexual reproduction takes place through the merging of a male cell (spermatozoid) with a female cell (ovum), leading to a fertilized egg. The germ cells are formed during meiosis, a particular form of cell division. Diversity of modes of nutrition: With the aid of chlorophyll, *autotrophs* make direct use of sunlight to synthesize sugars (mostly sucrose) from carbon dioxide and water. From these sugars, from mineral compounds contained in soils, and with water the organism synthesizes its own compounds, taking the necessary energy from the degradation of some of the sugars. The autotrophic organisms are the basis for all other nutritional cycles. *Heterotrophs* have to obtain their energy from the degradation of molecules formed by other organisms, animals or plants, but use some molecules to build up their own structure . *Saprophytes* live by absorbing certain compounds from decomposing animal or plant matter. *Osmiotrophs* take up the required complex molecules directly from the environment, filtering them through their body surface. These could have been the first organisms to originate in the "primordial soup" in which living matter formed (?). Autotrophic organisms appeared much later. The present classification increases the number of phyla by further subdivision of the *Monera.* (NB: The terms used here are defined in the Glossary.) (After Whittaker [1].)

become possible when a certain factual observation contradicts the generally held opinion and can no longer be explained by the current theory. This "unresolvable paradox" then requires a new hypothesis, theory, or model which, in the field in question, is then referred to as a new paradigm.

Theories are thus open to replacement, either by becoming invalidated by a subsequent theory or by integration in a new, more explicative synthesis, as is the case for Newtonian and relativistc mechanics. This also applies to the synthetic theory of evolution which, having first been formulated in the 1940s, should now be reassessed.

One tends to believe that science is perfectly objective, a perception that applies especially to the physical sciences which gain a certain authority from the mathe-

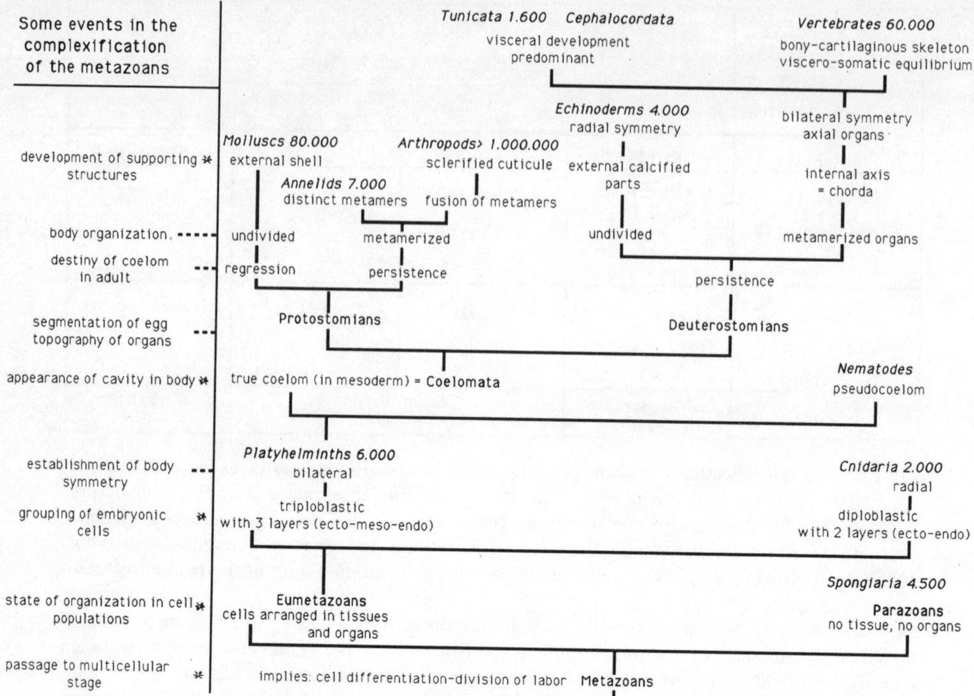

Fig.1.3. Some important stages in the increase in complexity of the metazoans. This figure is neither a classification scheme nor a phylogenetic tree. It should only serve as an illustration of the processes and patterns of increase in complexity in the metazoan world. Of the numerous phyla known, only those referred to in this book are shown. the temporal sequence of appearance of the various phyla is not strictly adhered to. As far as we know, the cnidaria and the (pro-arthropodas) were the first to appear about 680 million years ago, whereas the sponges, molluscs, and arthropods were already present at the start of the Cambrian (580 million years ago). The vertebrates, the youngest metazoan phylum, only made their appearance during the upper Cambrian (500 million years ago).

The lines between the phyla should not be understood as portraying an ancestor/descendant relationships as these are practically unknown and are the subject of much speculation. They should only serve as an indication of the paths of the changes connecting the various remarkable organic events such as the triploblastic stage, the appearance of the coelom, of the skeleton, etc. These events are documented by either fossil or living forms.

Increase in complexity is a well-established phenomenon in the metazoan world. It is a global, irregular, burgeoning tendency which, except in certain cases, does not provide an indication of the importance of a particular degree of organization. A vertebrate is evidently more complex than a sponge, but is a cephalopod mollusc more or less complex than an insect? Their body plans are different, but neither may be called superior (or inferior) to the other. The term "invertebrate", as opposed to "vertebrate", has not been retained, as invertebrate does not define a phylum, as in the case of the vertebrates, but rather a group of phyla charactcrized by the negative feature of the absence of vertebrate (internal skeleton). (NB: The terms used in this table are explained in the Glossary.)

* indicates that this event coincides with a makor evolutionary step, i.e. the extension of organic diversification. The causative relationship between such an event and others not marked with * is neither evident nor has it been elucidated in all cases

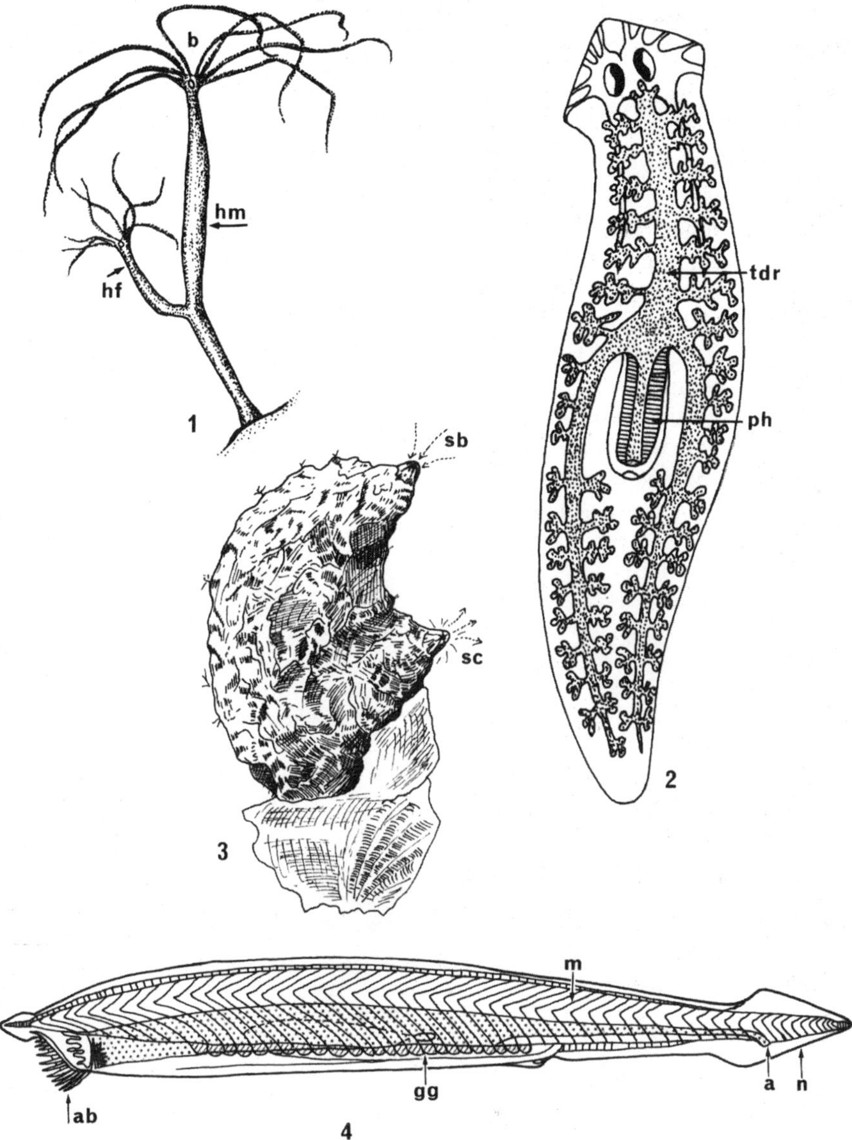

Fig.1.4. Some representative metazoans. *1* Freshwater hydra (cnidaria) multiplying by budding. *hm* Mother hydra; *hf* daughter hydra; *b* opening of the digestive cavity. *2* Freshwater flatworm *Dugesia* (plathyelminth). *ph* Pharynx pierced by the backward-directed mouth; *tdr* ramifying digestive tract. *3* The Mediterranean "violet", a tunicate or ascidian. Water enters through a siphon (*sb* mouth siphon) and is expelled through the cloacal siphon (*sc*), after having been filtered through an extensive pharynx that acts as a respiratory organ and as a structure for capturing food particles. *4* The cephalochordate *Amphioxus* or lancelet. External view showing the well metamerized muscles of the body wall. *a* anus; *ab* mouth appendages; *gg* genital glands; *m* muscles (myotomes or somites); *n* fin. (After Delage and Hérouard [10].)

matical data involved. This is, however, by no means true. Despite efforts to ward objectivity, science is also fraught with subjectivity and thus is not neutral at all. One of the reasons for this subjectivity, according to Hervé Barreau [8] is found in the interaction between science and the cultural, philosophical, and social factors of its time. At times, these factors appear even to control or obstruct science. Social factors have exerted much influence on the theory of evolution. the phyletic gradualism of the Darwinian theory, for example, was very much in accordance with the social philosophy of Victorian England which strongly believed in a gradual, progressive improvement of social conditions.

S. J. Gould and Niles Eldredge [9] specifically stressed this point by showing that their model of punctuated equilibria (see Chap. 11) conforms to the Marxist understanding of history. In this model, changes may take place only under pressure by revolutionary events or punctuations, the only factors able to break the stability and stagnation of the older social orders.

In conclusion, we want to point out that mysticism was a determining factor for the vitalistic and finalistic theory developed by Teilhard de Chardin, which in itself is a model of a metaphysical theory.

Chapter 2

The Modern Theory of Evolution: The Synthetic Theory

Some Stages in the History of Evolution

From his studies on *Invertebrate animals* and *fossils in the vicinity of Paris,* J.-B. de Lamarck (1744–1829) formulated a number of conclusions in his *Zoological Philosophy* [1] (1809), the first coherent theory of evolution. However, in contrast to commonly held belief, this was not based purely on speculation, but was progressively elaborated.

First, Lamarck established that life proceeded continuously throughout the history of the Earth. This principle of continuity of life contrasted markedly with the catastrophism of Georges Cuvier (in Laurent [2]), who postulated the extinction of successive faunas in the past and thereby a discontinuous history of the organic world. Lamarck could show that certain ancient mollusc species persisted unchanged in modern faunas, whereas others became modified to varying degrees. This clearly illustrated the continuity of life.

Secondly, he proposed the concept of transformation of species. Not only the dissilimarities between chronologically successive species, but the reasons for similarities had to be explained. Lamarck thus postulated the concept of evolving species. But in which way did they transform? This is where we enter the field of speculation.

For Lamarck, one mechanism of transformation was the desire of organisms to adapt to certain conditions of their environment. The utilization of a certain organ led to its development, whereas non-utilization resulted in disappearance. By heredity, the acquired features are then transmited from generation to generation.

None of these proposals were retained in modern science, despite sporadic neo-Lamarckist notions by palaeontologists such as Edward D. Cope. Molecular biology, however, has shown that the information (or the instructions) necessary for the construction of an organism proceeds only in one direction: from the genetic programme to the proteins. The appearance of a certain acquired character can not modify the programme, as this would necessitate a reverse transmission from the protein to the genetic programme.

It was Charles Darwin (1809–1882) who really introduced the concept of evolution to biological thought [3] where it remains firmly entrenched, despite certain reservations still maintained by some people. His ideas, as expressed in *The origin of species by means of natural selection,* were based on both observation and integration.

Ernst Mayr [4] has shown that the so-called Darwinian revolution is based on five facts and three deductions, to which we would like to add a number of observations.

First fact: All species possess a great potential fertility which would facilitate the rapid growth of successive populations if all members survived and reproduced.

Second fact: Except for annual or occasional fluctuations, the number of individuals in a population remains stable.

Third fact: Natural resources are limited and remain fairly constant in a relatively stable environment. "Resources" refers to the availability of space and nourishment.

First deduction: Theoretically, the number of individuals should rapidly surpass the capacity of the environment. This disequilibrium leads to a "struggle for survival" which allows only a small portion of the offspring to reproduce. Darwin used the term "struggle" in a metaphysical sense. Instead of this term, which suggest violent confrontation, the term competition is preferred, which does not necessarily imply a face-to-face encounter. Darwin stressed that this aspect concerns the individual members and not the species as a whole (see fact No. 4). The concept of competition was suggested by Darwin after reading Thomas R. Malthus [5]: "In the animal and plant kingdoms, nature has distributed the seeds of life with an extremely generous and liberal hand. Space and nourishment necessary for their growth, however, are rather rare... This implies that the difficulty in feeding results in a strong constraint constantly acting on the population... The races of plants and animals adhere to this grand law of restriction."

Fourth fact: No two individuals of a population are identical and the entire population exhibits a strong variability. This short formula explains some essential points of the Darwinian revolution. Pre-Darwinian authors like Linnaeus and G. Cuvier only used a species concept according to which type or model of a species reproduced immutably, the observed variations only representing "copying errors". For Darwin, variability was one of the fundamental properties of a species. There is no special type and all individuals are representative of their species to the same degree. It was this introduction of the concept of a population which was to subsequently play a profound role in biology.

Fifth fact: A large portion of this variability is inheritable. Variability is thus the source of evolutionary transformation.

Second deduction: Survival during competition is not purely a case of chance, but depends partly on the genetic constitution of the survivors. This unequal survival constitutes a process of natural selection. This statement should be seen in the right perspective as the part played by chance and/or genetic determinism in selection varies wide by between species. The two factors, however, always act concurrently.

Third deduction: The process of natural selection leads to a gradual change in the populations over generations, i.e. to an evolution producing new species. According to Lamarck, evolution is an inborn, regular feature, directed in a linear pattern from the simple to the complex, from the lower to the higher (scale of living beings). In contrast, Darwin saw production and diversification as the fundamental processes of evolution.

This led Darwin to the concept of the common ancestor of past and present forms which exhibit certain mutual similarities: the wolf, fox, and jackal would have been derived from the same common ancestor, which in turn would have a common ancestor together with the lion and tiger.

This concept of descendance enabled him to justify and explain hierarchical classification: By diversification, an ancestral species leads to a genus, a family and then to an order, as the descendants move away from the original design.

To the great dismay of many of his contemporaries, Darwin pronounced in 1872 [6] that man and the primates had a common ancestor *(The battle of the apes)*. Man is not a separate creation, but forms part of the entire organic world, reflecting an admittedly unique evolution (see Chap. 14).

At a later stage, August Weismann purged Darwinism of certain Lamarckian concepts by rejecting any possibility for transmission of acquired characters in a rather strict "Neo-Darwinism" [7]. Thus, only one mechanism remains: natural selection.

To explain the transmission of characters, Darwin used the theory of "blending inheritance", according to which the contributions of both procreators mix, somewhat like fluids. In this way one would obtain a progressive "dilution" of the variability with each successive generation and thereby achieve a uniformity between the individuals of a population. This would imply that the variability would eventually cease to exist. Had he known of the work of Gregor Mendel [8, 9], he would have been able to respond to his adversaries. The hereditary contribution of each parent behaves like "particles" (the future genes) which maintain their individuality over generations, joining for create certain features and to transmit them according to precise principles, the so-called Mendelian laws.

Hugo de Vries showed that in certain onagracean plants abrupt, spontaneous, and inheritable transformations occur which he called mutations. Around this observation, he built the mutationist theory of evolution, according to which new species appear abruptly, without intermediate stages (in contrast to Darwin). These new species are immediately stable and natural selection then removes all but the viable forms [7]. During the early 1900s, the applicability of Mendel's laws for plants and animals was rediscovered.

The study of hereditary transmission gained strong support at the beginning of the 20th century, in particular, from the work of Thomas H. Morgan and his school on the fruitfly, *Drosophila* [10]. From this work, we shall select particular aspects important for evolutionary mechanics. According to these authors, mutations are the sole source of transformation of organisms. Other potential causes like selection were not taken into account. This extreme, experimentalist position cannot satisfy zoologists and palaeontologists, as mutations, by their very nature, are random in occurrence and effects. Palaeontologists, as students of the fossil record, confirm that the organic world is not in chaos and that it undoubtedly exhibits directed transformations sustained over longer periods, the so-called lineages. This determinism should result from unknown factors and from natural selection.

It appeared that the two positions were incompatible with each other, and the study of evolution thus entered a crisis which was to be resolved only by the union of the two contrasting concepts in evolutionary genetics from the 1920s onwards [11].

These studies of evolutionary genetics of populations represent the joint effort of a large number of authors [12]. Ronald Fisher and John B. Haldane from Great Britain, and Sewall Wright from the USA, elaborated the theoretical models: Sergei Chetverikov from the USSR studied natural population of *Drosophila;* and in France, P. L'Héritier and G. Teissier carried out experiments on the genetics of individual populations in the laboratory.

Subsequently, Th. Dobzhansky and his school in the USA studied populations of *Drosophila* in the laboratory and in the field in Central and South America [13, 14].

As in Morgan's genetic studies, *Drosophila* was to become the choice material of evolutionary genetics, leading to the synthetic theory of evolution in the 1940s. This theory was influenced in particular by the systematic studies of E. Mayr, an ornithologist, and George G. Simpson [15, 16], who in his book, *Tempo and modes in evolution,* stressed the indispensable contribution of palaeontology to a coherent, explicative concept of evolution.

The genetic studies of Morgan were genealogical in as much as they start from an individual parent pair, tracing the transmission of certain genes through the line of successive descendants. Evolutionary genetics deals with populations: In a group of individuals making up an initial population, it establishes the frequency of a certain gene within the totality of the genetic programme of all individuals. It then goes on to study the potential variations of this frequency over successive generations. The source of all evolutionary change lies in the variations of frequency in genes throughout successive generations, controlled by external environmental factors, in particular by natural selection (Appendices 3.1 and 3.2) [17].

What Does the Synthetic Theory Tell Us?

Among the individuals of a popular or a species, the visible or hidden characters possess a hereditary variability which is the genetic component of evolutionary change. Within this variability, natural selection only retains what is useful to the organism. Furthermore, the permanent interaction between variation and selection ensures the adaptation of the organisms to their environment: adaptation is the motor of evolution. Eventually, the accumulation of small random variations screened by selection over generations leads to evolutionary change. A major transformation cannot take place in a single, large step as this would interrupt the adaptation of the organism and lead to its extinction. Small modifications occuring over different periods of time are at the root of all evolution, whatever its extent.

This will be our starting point to illustrate the present situation regarding the concepts of evolutionary mechanisms and to show the modifications necessary for their bearers, taking into account our increasing knowledge.

We immediately have to note one grave shortcoming of this synthesis [18]. It jumps directly from the genes to the resulting organism subjected to natural selection. In doing so, it ignores the whole array of growth processes, which form this organism according to the instructions (or programmes) of the genetic material. This omission forces us to admit that a certain gene is retained by selection more readily than another of lower selective value. Such a statement overlooks the fact that selection acts on the organism and is not aimed directly at its genes. A coherent concept of evolutionary mechanisms should satisfy the following scenario; genetic material – development – organism – descendant. Thus, selection affects development, the organism, and its descendant. Why development is not taken into account will be discussed later.

Artificial vs Natural Selection

The concept of natural selection was elaborated by Darwin from observations made by breeders practising artificial selection.

Artificial Selection: Man's Greatest Biological Experiment – from Wolvers to Dogs

From a wild source, the wolf, breeders managed to obtain some 300 different varieties of domestic dogs by practising artificial selection over a period of about 10 000 years since the Neolithic period. This clearly shows that the wild source, "wolf", contains a vast, hidden stock of variability, profitably exploited by man. It must have been possible to retain particular types of variation and eventually to accumulate and extend them over generations. This led to forms as extreme as the chihuahua (average height 20 cm; 1.8 kg) and the Great Dane (height 80 cm; 60 kg). Should these two dogs have been known only from fossils, they would have been interpreted at least as two separate species or possibly even as two distinct genera. We know, however, that they belong to the same species (Fig. 2.1) [19]. This procedure was not influenced by the environment or by selection, but entirely by the breeder's desire to develop the qualities demanded by man's particular needs: shepherd dogs, sleigh dogs, or dogs for pleasure. To achieve a certain desired result, the breeder selects by conserving in each following generation only those features which maintain or accentuate certain characteristics, and by eliminating others. In this principle of artificial selection, we recognize an operation directed toward a certain goal by human intelligence. From his interest in the different races of pigeons, Darwins was able to identify the methods and results of artificial selection, which he used as a basis for his arguments leading to the recognition of natural selection.

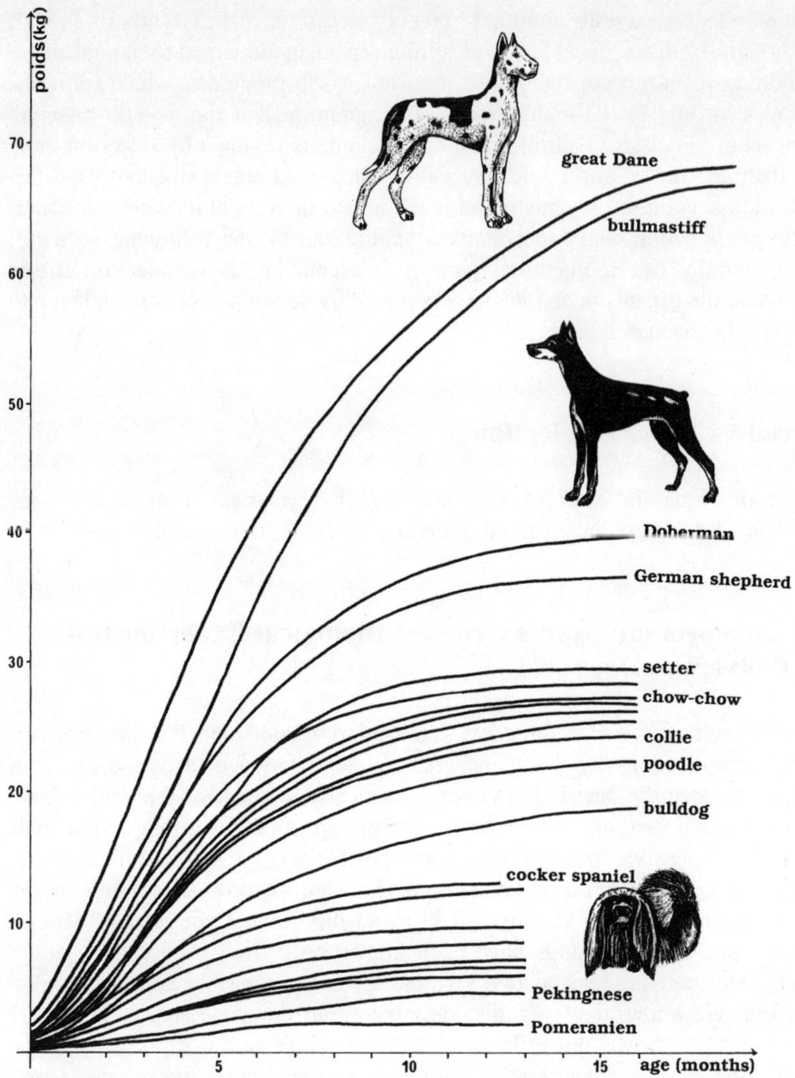

Fig.2.1. Man's greatest biological experiment. The graph shows the weight vs. age for 22 breeds of dog. That of the wolf would fall between the Doberman and the bullmastiff. (After Wolter [19].)

How Do Natural and Artificial Selection Differ?

For natural populations we have to take environmental conditions, the phenomenon of competition, and the immense duration of geologic time into consideration and we have to reject any idea of predetermination.

Natural Selection at Work: Competition

All environments are limited in their extent and resources. Thus, the various organisms occupying a particular environment have to enter into competition with each other for space: The hunting territories of a lion and a buzzard cover several square kilometres, while that of a field mouse covers only a few square metres. Thus, in a given territory, more field mice can live than lions or buzzards.

An other aspect of competition is food: Plant resources are much more numerous than animal resources and, as a result, the frequency of herbivores or omnivores is much higher than that of carnivores. The African antelopes and zebras, for example, occur in large herds, whereas lions only form small groups or live solitarily.

Competition may occur between individuals of the same species or between different species, leading eventually to selection. This selection will lead to different rates of reproduction, resulting in at least three processes: survival, success in reproduction, and differential fertility. These processes, however, do not represent the obligatory outcome of the limited nature of the resources and thus of competition. The phenomena of selection and competition are not necessarily linked. Malthus, for example, recognized competition as a factor, but was unaware of selection.

Natural selection is the result of the very conditions of existence. It is not controlled by intellience, but acts, as it were, from "case to case". It does not "know" the future of a species and cannot operate on the basis of "long-term benefits".

Natural selection is not just the brainchild of human imagination like many of the supposed "forces" of evolution. It is an observable fact and a reality recognized under natural and experimental conditions. Natural selection controls and directs evolution by acting on the variability of the entire population. In contrast to the statements of Darwin's adversaries, its role is not limited to the elimination of the weaker or less fit individuals ("selection of the weak") or, in general terms, of anything outside the norm. If this were its only effect, it would be a force of conservation rather than change.

Let us assume that from one generation to the next, selection tends to retain larger indidivuals at the expense of others. The mean size of the species will then gradually shift to increasingly larger values under the influence of directed selection, the effects of which are readily observed. One can thus imagine that instead of retaining only individuals of average size, weight, colour, etc., selection tends to preferentially eliminate them. Only the forms deviating more from the average will survive and the population (or species) will eventually find itself divided into two. A multiplication of the number of species will result from this diversifying (or disruptive) selection which, however, is more theoretical compared with the other two types of selection. It is primarily deduced a posteriori and not directly observed. When an ancestral species is divided within a lineage into two descendant species, diversifying selection is frequently invoked to explain this divergence.

Natural selection is also reffered to as a "struggle for survival", a term used by Darwin. However, "I should premise that I use the term Struggle for Existence in a large and metaphorical sense, including dependence of one being on another, and including (which is more important) not only the life of the individual, but success in leaving progeny. Two carnivorous animals in a time of dearth, may be truely said to struggle with each other which shall get food and live. But a plant on the edge of a desert is said to struggle for life against the drought, though more properly it should be said to be dependent on the moisture" [3].

It would be better to abandon the notion of "struggle", as this has resulted in numerous misinterpretations by authors wanting to justify war or social inequalities.

It must also be stressed that competition does not necessarily imply a direct confrontation. Competition can also take place via a third party. In the 18th century, James Cook introduced goats to the Galapagos Islands to allow ships to restock with fresh meat. The goats proliferated and, at the same time, the number of giant tortoises decreased. Competition for the same food without direct confrontation of the two participants was the cause of this phenomenon.

Selection and Chance in the Survival of Organisms

At the very start of its life, a young lion may succumb to sickness or an accident. Once an adult, it will have no natural enemies and only its individual qualities, through survival, will ensure its capacity to perpetuate the existence of the species. In such a life cycle, natural selection plays a decisive, but not exclusive, role.

Shrimps of the genus *Euphausia* constitute much of the plankton fed upon by whales. Once a whale has swallowed several cubic metres of seawater teeming with plankton, it filters it by expelling it through the baleen. The destiny of an individual shrimp will thus not depend on any of its own qualities: Survival or death depends only on chance, and selection will favour only one quality, the higher or lower rate of fertility of the various populations of *Euphausia*.

A herring produces thousands of eggs which whitings swallow in great quantities. The fry fall prey to predators and the number which escape such obstacles, determined by chance, to become adults is rather small. Even those surviving are still harassed by predators and the survival of the species rest on only a very few specimens. In this enormous mass, in which chance prevails, selection can only favour reproductive capacity.

There are ample examples from other groups of animals, such as parasites like the liver fluke or Taenia. The success of their survival is controlled by chance and their fertility. From this we can draw one conclusion: natural selection and chance both control, albeit to varying extents, the destiny of individuals and species. No life cycle depends exclusively on chance or on selection. There are no general rules and each cycle should be analysed separately. Where several populations confront each other in the same environment, we are not able to predict how competition will take place and what its outcome will be. It will only be with hindsight that we can evaluate the results and assign a higher selective value to one of the contestants. This is the conclusion drawn by palaeontologists and zoologists alike: natural selection has no predictive value whatsoever.

Except for some rare cases, we are also unable to recognize the important factors in competition, even between living forms, although all the information we could wish for is available. Why has the portuguese oyster been able to virtually wipe out the indigenous flat oyster, restricting it to only a few protected sites? Why is it that the grey rat, immigrating to Europe from Asia in the 18th century, managed to confine the black rat, the original occupant, in particular to rural habitats. In palaeontology, where only partial information is avalaible, how can we determine the factors decisive in competition from only a shell or a skeleton? In the lineage of the Tertiary horses, *Mesohippus,* which flourished in the Oligocene [20}, was replaced by *Merychippus* at the start of the Miocene (Fig. 2.2). We can only acknowledge the fact. Was the former less well-built than the latter, less efficient in making full use of the environment, or was it slower? We just do not and cannot

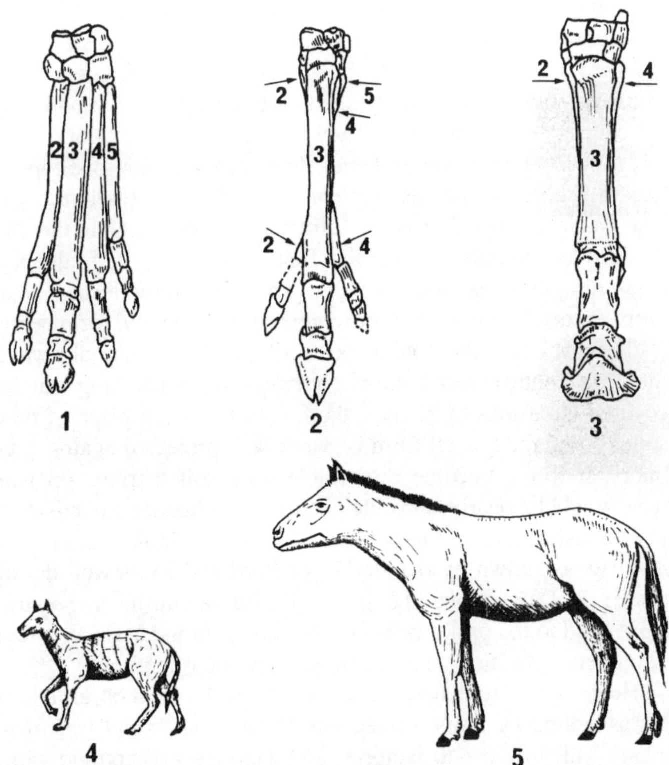

Fig.2.2. Fossil horses: stages in the reduction of the number of digits. *1* The Eocene *Hyracotherium,* front leg with four digits (the foot itself consists of only three toes). *2* Leg of the Miocene *Merychippus.* The central toe (No. *3*) has become the most important one, supporting virtually the entire weight of the body. The lateral toies (Nos. *2* and *4*) still possess three phalanges. *3* Leg of the modern horse (*Equus*) with only one toe (No. *3*); the lateral toes are reduced to splint bones without phalanges and are attached to the back of the principal toe. *4* Early Tertiary greyhound-sized *Hyracotherium. 5* The donkey-sized *Merychippus.* (After Simpson [20].)

know! Was there even confrontation between the two, leading to the elimination of one of them?

We have to recognize that, at least in some cases, the occurrence of competition-selection between slightly different forms is an assumption which cannot be verified in the fossil world. Moreover, another point has to be stressed: competition will not necessarily result in the complete elimination of one of the competitors. Frequently, a new, stable equilibrium is established due to the redistribution of territories and resources.

Natural Selection at Work Among Moths

In his *Origin of species,* Darwin did not *establish* natural selection on the basis of observations in the wild. He only *deduced* its existence from the ideas of Malthus (competition) and the experience of breeders (artificial selection). The concept of natural selection facilitates the *interpretation* of real and theoretical changes in modern and fossil faunas.

However, are there cases where selection may be observed or analysed while acting on natural populations? Well-documented cases are rare, and the best is represented by the peppered moth *(Biston betularia)* of Great Britain. This rather unobtrusive, black and white winged moth occupies an important place in the history of science, as it allows us to study natural selection at work.

Until 1849 collectors in Great Britain caught only one type of this moth, which was to become known as "*typical*". Then a black (melanic) specimen was caught near Manchester; this was the form *carbonaria* which in the 20th century was recognized as a mutation. After this first appearance, the frequency of *carbonaria* increased rapidly, reaching a value of about 90% in industrial areas by the end of the 19th century. In rural areas, however, only the *typical* form persisted.

To what could this change be ascribed? The industrial revolution started during the 19th century and factory chimneys were belching out black smoke which covered the trunks of birches, the favourite resting place of these moths. On a unsoiled trunk the typical form is rather well protected against predators through mimicry. It would become discernible on a soiled trunk, on which *carbonaria,* in turn, would be nearly invisible due to its "industrial melanism". It was thus natural to established a causal relationship between pollution and the proliferation of *carnonaria,* as shown by the studies of Ford and Kettlewell during the 20th century (see H. B. D. Kettlewell [21]). A selection favouring *carbonaria* at the expense of *typical* led to the proliferation of the former in industrial environments. In rural environments selection operated in the opposite direction.

How did this selection occur? As shown by field observations and experiments, it was caused by birds. Mixed lots of *carbonaria* and *typical* were set free in the countryside or close to factories and were later recaptured with the aid of ultraviolet lamps. Counts of the recaptured populations were self-explanatory: In the industrial zones there was a virtual massacre among the typical form, whereas *carbonaria,* although not entirely spared, suffered smaller losses. The birds initially caught the easily visible prey, but then learned to also spot and catch the camouflaged moths.

We are dealing here only with changes in frequency within a species and not with the process of speciation. However, this observation of natural selection at work enables us to draw some interesting conclusions regarding evolutionary mechanisms.

Melanism as a mutation not only affects this moth, but also some 70 other species of moths, one beetle, one millipede, etc. It is not restricted to Great Britain, but is also found in other extensively polluted areas, e.g. the Ruhr of Germany or in the Pittsburgh area (USA). The history of this moth is thus not a localized incident caused by a single random factor, but a phenomenon controlled by the sequence: pollution – selection – melanism, the correlation of which was confirmed by the reverse process. More recently the antipollution measures around Pittsburgh have started to take effect and the frequency of *carbonaria* has diminished rapidly.

The data gathered in Great Britain, furthermore, show that the evolution of this mutation takes place much more rapidly than expected from calculations of evolutionary genetics. It appears therefore that the selective advantage of the *carbonaria* mutation is much higher than considered theoretically possible by some authors. We are dealing here with a unique acceleration of gradualism (Chap. 11).

Part II
The Message of Nature Today – The Evolutionary Mechanisms

Chapter 3

Genetic Material and Programmes

The Structure of the Genetic Material: DNA and Genes [1, 2, 3, 4]

The genetic material containts the instructions necessary for the construction of an organism and ensures the continuity of the species by transmission of the construction plan from generation to generation over considerable periods of time.

This definition of the role of the genetic material, simple and rigid as it may be, leads to two questions: Does the construction of an organism rely exclusively on the genetic material? We shall discuss this in Chapter 4. Is the construction plan transmitted unchanged over generations? Certainly not, otherwise the species would be immutable and there would be no evolution.

The genetic material thus can, and must, change. This is proven by the everyday observation that no two individuals within a species are exactly identical, except identical twins from the same egg.

For example, the dog shows us that the genetic material, and with it the entire animal, may vary, at least within limits, so that the species "dog" is not transformed into another species.

Let us go back in time to the wolf, the ancestor of the dog, and to its own ancestor. Over millions of years in the Tertiary, there was a succession of species in an evolving lineage which eventually led to the wolf. Throughout this long history, the genetic material was thus not fixed, but varied to the extent that one species transformed to the next. This "unfaithfulness" in transmission is one of the reasons for evolution.

The genetic material of the chromosomes in the cell nucleus (Appendix 3.1) is made up of deoxyribonucleic acid, DNA [1] (Fig. 3.1). Functionally, the genetic material is divided into genes, and we shall state here provisionally that a certain gene is responsible for a certain function (Fig. 3.2).

Ribonucleic acid (RNA) is made up of a chain of nucleotides in which ribose occupies the position of the sugar and thymine (T) is replaced by uracil (U). Three letters of the messenger RNA, reffered to as a codon, define one of the 20 amino acids making up the proteins of an organism.

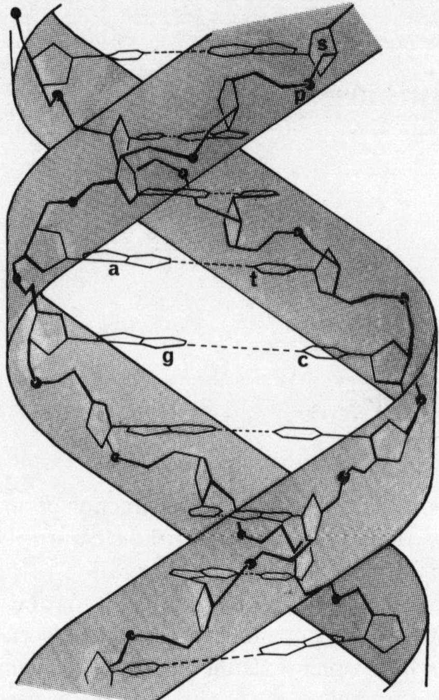

Fig.3.1. The DNA molecule. The DNA double helix consists of two intertwined strands, each made up of a band of sugar (*s*) and phosphate (*f*) molecules connected by strong bonds (*solid lines*). Each molecule of sugar (ribose) conists of five carbon atoms and is connected to two phosphate groups. To each sugar, an organic base is attached, facing to the interior of the helix. This base is either adenine (*a*), thymine (*t*), cytosine (*c*), or guanine (g). the bases from opposing band positions are connected by weak hydrogen bonds (*dotted lines*) according to a "complementary pairing rule": (*a*) only links with (*t*), and (*c*) only with (*g*). This sequence of bases constitutes the genetic code carried by the molecule.

How Can the Genetic Material Change?

The variability of the genetic material is caused by two possible factors: mutations and recombinations. The term "mutation" describes a change in the genetic material or the result of this change on the organism itself. At the gene level, mutations will modify the molecular composition. Chromosomal mutations affect whole groups of genes which may then disappear, multiply, or change their position on the chromosome. They may also pass from one chromosome to the next, and the number of chromosomes may also vary (Appendix 3.2, Fig. 5.5).

"Recombination" is a very important process for the variability of the genetic material. An egg cell contains only half the number (n) of chromosomes of other cells. On fertilization, the spermatozoon contributes the paternal "half" of the chromosomes to the egg cell, and the normal number (2n) of chromosomes is re-estab-

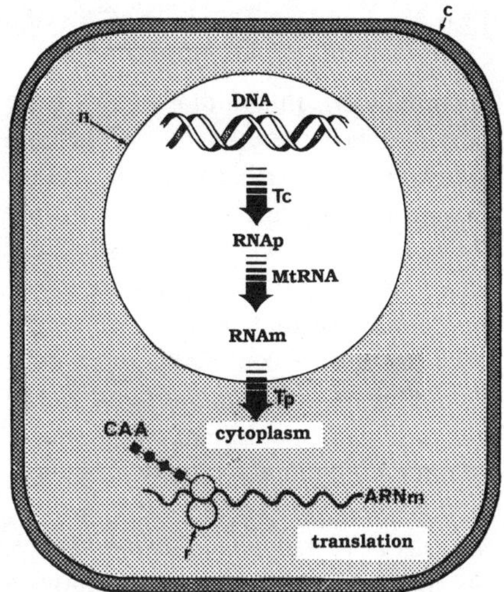

Fig. 3.2. Synthesis of a protein. In the cell nucleus the gene, defined by its *DNA*, is first transcribed (*Tc*) into primary or messenger RNA (*RNAp*) which is complementary to part of the DNA strand. During the process of maturation (*MtRNA*) parts which correspond to the (non-coding) introns of the gene are cut out from this RNAp. The sections corresponding to the (coding) exons merge to form the messenger RNA (*RNAm*) which leaves the nucleus through pores in themembrane (*n*) to enter the cytoplasm where translation into a protein takes place. Acting like a "reading head", a ribosome (*r*) attaches itself and reads the codons (groups of three nucleotides of the messenger RNA) one by one. Each codon represents one amino acid to which an RNA molecule attaches itself, the socalled transfer RNA (RNAt, not shown). These amino acids then assemble to form a polypeptide chain (*CAA*) which subsequently detaches itself from the messenger.

lished. The appearance of sexual reproduction was certainly a major event in evolution beyond the bacterial stage.

In conclusion, the genetic material possesses a considerable, if not gigantic, capacity for variation. Except for identical twins there are not two individuals with the same genetic make up throughout the history of a species. These variations in patrimony result in modifications of the respective organisms, but the organisms themselves still belong to the same species (cf. the dogs). They thus possess largely the same selective value in relation to the environment, except for serious anomalies like mongolism among humans.

We are now left with two questions:

Firstly, what genetic changes can cause the development of one species from another? We shall try to answer this in Chapter 7. Secondly, to what extent will the number of mutations (rate of mutation) control the possibilities of evolution, or more precisely, the rate of evolution of organisms? Different answers are given to this question by different authors (Appendix 3.3).

The Neutralist Theory of Evolution

According to some authors, most gene mutations are neutral or near-neutral, i.e. they have no repercussions on the structure or functioning of the organism. This approach is the basis for the neutralist theory of evolution as proposed by Motoo Kimura [5], derogatively referred to as a non-Darwinian theory of evolution by Thomas H. King and Jack L. Jukes [6]. From theoretical considerations and calculations too extensive to be reported here, these authors showed that, taking the observed rates of mutation into account, the organisms would not be able to withstand this nearly continuous pressure for adaptation, if every mutation had an effect on the organism and was subject to selection. For the most part, the mutations thus have to be neutral and, consequently, variations in gene frequency (Appendix 3.4) in the populations are not exposed to selection at all or, if they are then only slightly. According to Kimura, this neutralism only applies at the molecular level, whereas at higher levels of the organism, selection does indeed take place in all possible ways; however the mutations are not able to explain the transformation of the species. The neutralist theory made a considerable impact and led to the division of evolutionists into two opposing camps (Appendix 3.5).

Mutation and Chance

The acceptance that mutations occur at random signifies that within a population we are not able to predict which genetic material will mutate, which gene will be affected, when it will be affected, or what the effect of the mutation on the organism will be. Three of these questions concern the random nature of the mutation. The fourth only represents the fact that we are unable to analyse the chain of reactions between the genetic event and its translation into the organism, save for rare cases like abnormal hemoglobins. This unpredictability is not of an intrinsic nature, but results from our present techniques of investigation.

However, the effect of the mutation will be random in relation to the destiny of the individual, as the change has no connection to the adaptive requirements of the species in question. As the organic world is not in chaos, the enormous possibilities for random changes at the genome level must someshow be controlled and constrained. One of the constraints acting from the outside on the individual is natural selection. Although omnipresent, is this really the overpowering and sole directive force of evolution? We shall return to this question later.

For *Drosophila* and the mouse, the majority of the mutations will be unfavourable or even lethal. Is it then possible that mutations are the source of evolutionary change? To say that a certain mutation is useful, useless, or neutral does not make sense, as the conditions under which it occurs have to be defined. For *Drosophila,* wether a mutation (e.g. reduction of the wings) is favourable or unfavourable for the survival of the species [7, 8, 9] (Fig. 3.3) will depend on the specific circumstances. It has to be borne in mind also that a certain gene is not independent, as its influence is controlled to some extent by neighbouring genes. Theoretically, there

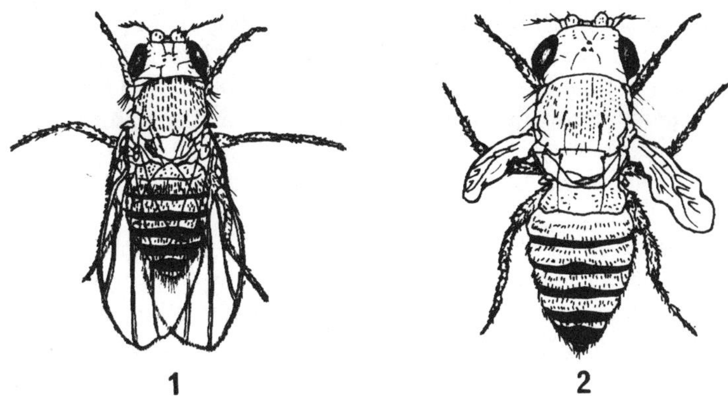

Fig.3.3. A normal *Drosophila* and the "vestigial" mutant. **1** Normal *Drosophila*. **2** *Drosophila* with "vestigial" (vg) mutation. (After Ayala [18].)

is thus no reason why a gene that is unfavourable in combination with certain other genes should not become neutral or even favourable in a different genetic "neighbourhood".

Functioning of the Genetic Material

When the establishment of a certain feature begins, there is cooperation between a pair of genes. However, what is the nature of this connection between the activity of the genes and the construction of the organism? Let us look at the reduction of the wings in *Drosophila* (Fig. 3.3). We note that this spectacular is accompanied by a number of smaller changes in other structures which do not exhibit any anatomical or functional connection with the wings. This influence of one gene on several independent features is referred to as pleiotropism.

When gene vg + mutates into its allele vg, the results is unfavourable in the normal living conditions of the fly and thus is not retained by selection. However, when a population of flies with normal and vestigial wings is exposed to a strong wind in a cage [7], those with vestigial wings are favoured as, when sitting on the walls, they are not carried away by the wind as readily as the normaly winged flies. The loss of wings through mutation is thus not always unfavourable for insects, as quite a few species, living on islands and on the continents, have experienced such a loss without adverse effects [8, 9].

Certain transformations do not appear to have repercussions on the vitality of the mutant. However, there are others that do have a profound influence on survival: reduced fertility or reduced larval vitality. From this observation we can imafine an evolutionary scenario: in a wild population of *Drosophila*, the environmental conditions will favour the reduction of the wings. However, this adaptation cannot establish itself completely in the population, as it is counterbalanced by unvavourable effects of this mutation on fertility and larval vitality.

Pleiotropism certainly plays an important role in evolutionary change. It is thus dangerous to come to premature conclusions on the reasons for success or failure of competition, or for the extinction or expansion of a species, present or past, based only on the limited observations of teeth, bones, or shells. We would be ignoring the rest of the organism concerned.

In another phenomenon, polygeny, several genes combine their action to form a measurable feature. The hypothesis of polygeny was advanced to explain the size variations in maize: to a basic height common to all maize plants, each gene of a polygenic complex may add a few centimetres. The heigh is thus a function of the number of active genes. The same mechanism is invoked to explain, for example, variations in pesticide resistance within a population of a certain insect species.

Polygeny could explain the phenomenon of a change in size, so frequently found in evolutionary lineages.

Genetic Programmes in the Synthetic Theory

The founders of evolutionary genetics [10] have assumed a rather simple natur for the genetic programme by visualizing the genes as being lined up like pearls on a string, from which they may be removed completely independently of one another. To each gene a selective value may be attributed which is stable over many generations.

The synthetic theory stresses the variation of gene frequency within the entire genetic programme of a population from generation to generation as the mechanism of all evolutionary change (Appendix 3.4). In addition to mutations and recombinations, the frequency variations result from another random phenomenon, namely genetic drift.

In an effectively constrained population, an allele may be either eliminated (frequency 0) or retained (frequency 1). Its destiny thus does not depend on its selective value, but represents the consequence of a sampling procedure of small quantities. The validity of this phenomenon of drift can be demonstrated mathematically, but it is still disputed by certain authors. It is now taken into account in models of speciation such as the "founder population" (Chap. 7).

Is this rather simple process of variation in gene frequency able to explain gradual evolution, one of the main pillars of Darwinism and of the synthetic theory? So far, no entirely satisfactory answer to this question has been proposed. Is then one single type of evolutionary mechanism responsible for all changes? Or are there two different mechanisms, one for microevolution within a species, and another for macroevolution, responsible for the diversity of the various species?

In the microevolutionary concept any change, of whatever extent, may take place only by the cumulative effect over time of minimal transformations which are all selected the same way. This would imply that the same gene or group of genes always has to mutate in the same sense in multiple steps and over long periods. As mutations occur at random, the evolutionary change would result from a lottery in which the same number always won. The same number may recur twice or even three times, but with an increasing number of draws it would become increasingly improbable and eventually mathematically impossible. This objection against gradual evolution has been raised repeatedly.

Over a period of 10 years, the number of generations involved amounts to several tens of thousands for a bacterium, some hundreds for *Drosophila,* between 50 and 100 for insects and small mammals, but reaches only a few generations for elephants or man. If evolution depended only on the possibility of repetition and accumulation of mutations with time, bacteria would posses a much higher evolutionary potential than elephants and man. However, such a conclusion is rendered invalid by the facts. During the 60 million years of the Tertiary, elephants have shown a truely remarkable diversification. And human evolution is one of the most rapid known (cf. Part Four).

The "New" Genetics

The results of molecular biology have profoundly reorientated our understanding of the structure and functioning of genetic programmes. However, what are the repercussions on the problems of evolutionary mechanisms?

Progress in molecular biology has been rapid indeed, but only a few authors ask themselves about the potential ramifications of this progress in the field of evolution [12, 13]. Of the wealth of new data, we shall discuss here only those directly applicable to our problems. The bacterial gene – the object of the molecular revolution – is of a compact type, i.e. all its DNA codes for proteins and genes are linked together. In contrast to this, the genes of other organisms are of a particulate type in that the coding portions (exons) are separated from each other by non-coding regions (introns), the role of which is still poorly understood.

This organization increases remarkably the capacity for expression. Translated as an entity a gene may code for a certain protein, whereas in a divided state, each portion of the gene may result in a different protein. It is also possible that two different genes could combine to produce another protein with different properties. This increased potential for expression with a limited quantity of DNA leads to an economic utilization of the genetic programme. As a consequence, a complex organism like an insect or a mammal will not require a considerably larger amount of genetic material than a bacterium, as the difference in complexity would suggest. The coli bacillus *(Escherichia coli)* requires some 3,000 genes compared with possibly only 100,000 genes necessary for man.

By repeated multiplication [3], an ancestral gene leads to a family of multigenes, all members of which exert the same, similar, or totally different functions. Such families of genes result in a highly economical use of the genetic information. Among the families of multigenes responsible for the formation of antibodies, which are produced against outside attack by antigens, a limited number of genes are able to initiate the innumerable immediate responses that are required by the organism for survival during its entire life span.

Immune reactions are a good, and only recently discovered, example of this capacity to react immediately in an adaptive manner to an environmental demand or a sudden selective pressure. There is thus no need to call up a "miraculous" mutation or a sluggish gradual transformation. This manifestation evidently cannot be gleaned from the palaeontological record, but there is no reason to doubt the action of comparable mechanisms on other types of structures.

The "new" genetic programme is no longer considered to be rather rigid and only able to be transformed by mutations and recombinations. Mobile elements, also referred to as jumping genes, may spontaneously relocate from one position in the genome to another, thereby modifying its activity in a sometimes radical fashion. Jumping genes, first discovered in maize by B. McClintock (see Lin Chao et al. [14], have now been identified in a variety of organisms. A virus may insert its own genetic material into the genetic programme of the host and affect the host's functioning. Authors like Francois Gros [1] tend to assign a profound influence to these jumping genes in the process of specication. This interesting approach has recently gained much attention [13].

A final aspect of the "new" genetics, which could lead to a new understanding of evolutionary mechanisms, concerns the hierarchical functional organization of the genetic programme. As we have outlined, in the synthetic theory the genetic programme represents a linear system of instructions based on structural genes which encode the proteins required for the construction and functioning of the cells. Modern research, however has revealed the existence of categories of genes which directly affect the function of other genes. These are the regulatory genes which control the activity of structural genes. This new understanding of the genetic material as being subdivided into several hierarchical functional levels may be compared to the functioning of the brain, in which the so-called higher centres control the primary centres for perception and motion.

The famous "operon model" of Francois Jacob and Jacques Monod (see F. Gros [1]) has shown that the function of bacterial genetic material is regulated by signals which may result from its own activity or from substances in the surrounding environment. In higher organisms such signals are enhanced by other substances (such as hormones); these are produced during embryonic development and are subsequently recognized by regulator genes. The presence of these regulator genes is now well established, but their actual organization in the hierarchical functional system is still subject to speculation [15] as is shown schematically in Fig. 3.4.

This new type of organization allows us to envisage a plausible mechanism for gradual evolution: According to the synthetic theory, in the genetic programme, which is made up of individual structural genes, a group of genes that is responsible for the execution of a certain feature would mutate spontaneously several times and always in the same direction so that small transformations accumulate. With this organization of several functionally interconnected levels, the effects of the activation of a group of structural genes would not be restricted to one level, but could be initiated by numerous other genes within the network. In this way, the possibility of obtaining a repetitive effect with time is greatly enhanced.

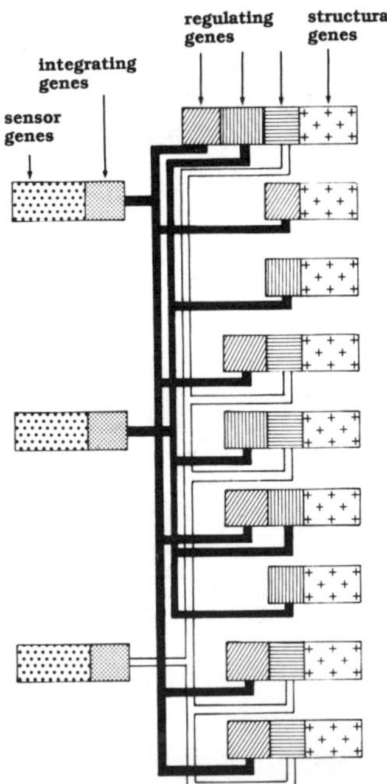

Fig.3.4. Model of the functional organization of the genome of higher organisms. The *black* and *white lines* illustrate the potential relationships between the different functional stages of the various genes (a representation by Valentine and Campbell [16] of the model proposed by Davidson and Britten).

Within this complex network a gene will not stand by itself, almost independent and free "to act at will" like the genes in the models of evolutionary genetics. As a member of an organized society cooperating with other genes, it will be subjected to the rules and constraints of this organization. Here we observe for the first time the concept of internal constraints, which we shall encounter again when considering other stages in the construction of an organism. This idea leads to a radical modification of our concept of the evolution of organisms.

Some final words on the processes involved in the evolution of genetic material from the simplest to the most complex beings: The genome of a mollusc, an insect, or a vertebrate should contain more information than that of a bacterium. Does this simply require a increase in the quantity of DNA? It does to some extent, as shown by the corresponding number of genes. However, another aspect might be more important for the increase in capacity of the genetic programme during evolution: One bacterial gene encodes one protein, two genes code for two proteins, etc. but two genes in a higher organism may code for several proteins. This economic use of genetic material and information is linked to the establishment of increasingly more elaborate systems of regulation and coordination. We shall also see that important transformations in organisms may be initiated by a small number of genes or even by a single gene.

Concerning the growth of the genetic programme during evolution, the development of regulating networks and the redistribution of gene activities will also certainly have played a more decisive role than a mere increase in the quantity of DNA.

Molecular Evolution

The classical morphological data of comparative anatomy have recently been supported by analytical results on the proteins, DNA, and RNA of living organisms [17, 18] (figs. 3.5 and 3.6).

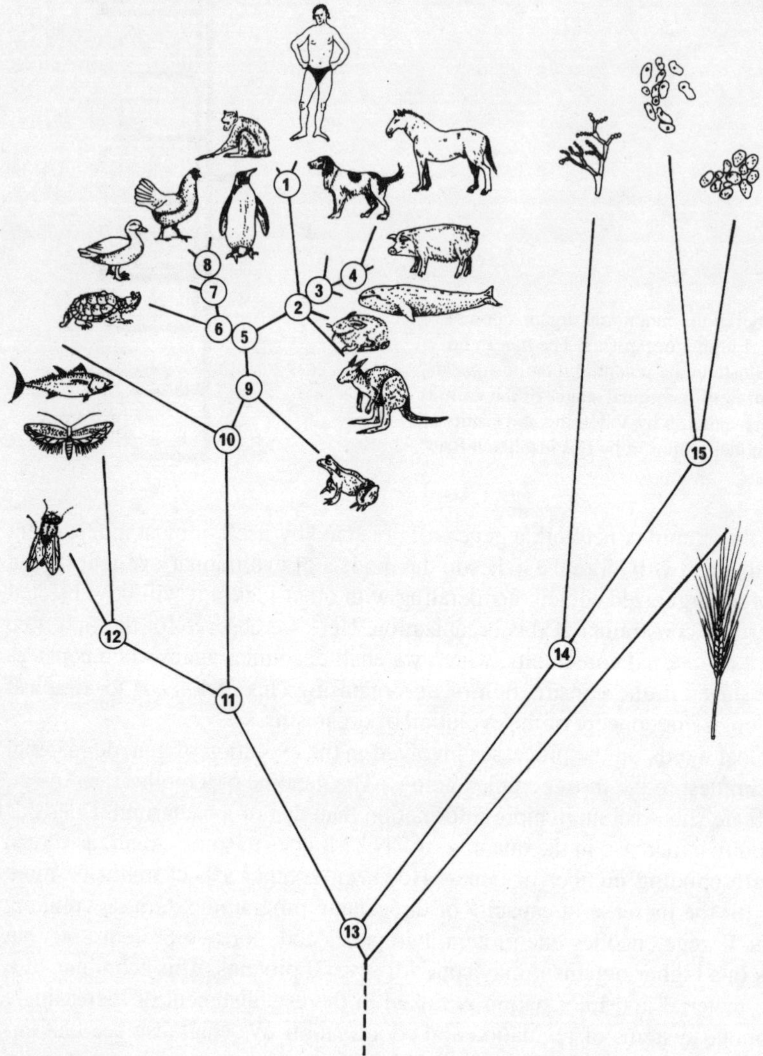

Fig.3.5. Molecular phylogenetic tree based on cytochrome C. The amino acid sequence of this protein of the respiratory chain was analysed in different plant and animal organisms, and a computer programme then established the possible sequence of ancestral proteins and the number of mutations (not shown) which would be necessary to progress from the ancestral to the derived structure. From structure *10*, for example, there were 11 mutations to obtain the tuna cytochrome, whereas there were 3 mutations to reach structure *9* from which the cytochrome of the frog was derived as well as the ancestral structure *5*, etc. (After Ayala [18].)

When comparing the amino acid sequence of cytochrome C, a protein of the respiratory chain, between the rhesus monkey and man, we note that of the 104 amino acids, 103 show the same order in both cases. The one exception is position 66 which is occupied by isoleucine in the rhesus monkey and by threonine in man. During the evolutionary divergence separating the two species, only this one substitution of an amino acid took place, whereas 12 substitutions separate man from the horse and 11 substitutions occurred between the horse and the rhesus monkey.

If we assume that the number of substitutions increases with time in a regular pattern, type of molecular clock, it would be possible to estimate the relative time that has elapsed since the onset of divergence between different lineages and to construct a moleclar phylogenetic tree from these data (Fig. 3.5).

Compared with morphological methods, the molecular technique allows a comparison between fairly different forms which do not exhibit common features, such as horses and butterflies or yeasts, etc. Molecular evolution may also take place through the doubling of an ancestral gene, leading to the establishment of different lineages of proteins in the same chemical family (Fig. 3.6).

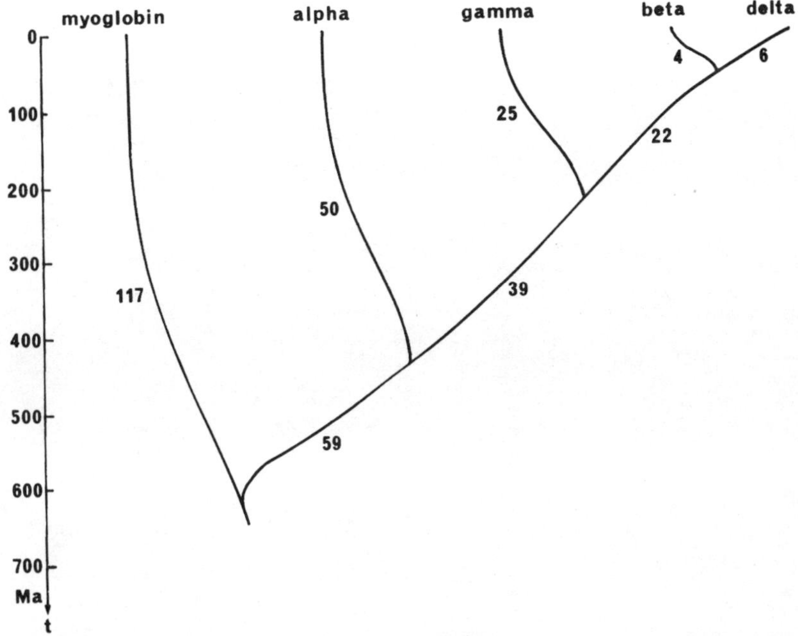

Fig.3.6. Genealogical tree of different globins. The muscle myoglobin and the blood haemoglobin are the transporting agents of oxygen. The divergent points indicate the stage at which the ancestral genes became separated, giving rise to a new line. The minimum number of nucleotide replacements required to explain the differences in amino acids between the different proteins is shown along each branch. The first gene divergence occurred some 600 million years ago, resulting in a gene for myoglobin and an ancestral gene for the various types of haemoglobin. About 400 million years ago, this haemoglobin gene split into gene alpha and another gene which itself was split 200 million years ago into the forms gamma and beta. From gene beta, some 40 million years ago, the ancestral line for the higher primates diverged, giving rise to a new gene encoding the delta chain of haemoglobin. (After Ayala [18].)

The proteins, and with them the genes, are not all modified at the same rate. In cytochrome C, the rate of subtitution is very low, whereas for haemoglobin it is medium, and in the fibrinopeptides, proteins appearing during the coaggulation of blood, it is very high. The rates depend on the role of the respective molecules in the functioning of the organism in question.

One problem in molecular evolution is the reliability of the clock. That the substitutions take place at regular time intervals is a useful assumption, but the authors are aware of the fact that the movement of this clock does not possess the regularity of a time control. However, despite these "fantasies", the clocks of various substances do provide coherent and useful results over longer periods of time.

Once the date of one of the many bifurcations in a lineage is known from absolute (radiometric) data, it becomes possible, by knowing the number of substitutions that have occurred since this point, to calibrate the clock in absolute terms.

Chapter 4

From Egg to Adult: Development

Relationship Between Development of the Individual and Its Evolutionary History

We may state that the path of development of an organism can play an important role in its immediate realization and also in its possible transformation. This leads to the problem of the existence of a causative link between individual development (ontogeny) and evolutionary history (phylogeny). This problem which has been dealt with by numerous authors, has led to contradictory answers, depending on the direction of the supposed connection: from the evolutionary history to individual development or vice versa, that is, from individual development to the evolutionary history.

Is Development the Resumé of Evolutionary History?

Between 1866 and 1874, Ernst Haeckel [1, 2] wrote his *Fundamental Biogenetic Law* which soon became famous and which, even today, continues to influence evolutionary thinking. This "law" states that "ontogeny recapitulates phylogeny", i.e. the history of a lineage appears in a condensed form in the development of its present descendants. The law then goes on to pronounce that "phylogeny is the mechanical cause of ontogeny". The first of these statements permits us, in principle, to reconstruct the evolutionary history of recent forms without knowing, or by ignoring, their fossil ancestors. This aspect of Haeckel's "law" is the best known and used most often.

Let us take a look at a classical example and compare the very young embryos of a shark, a lizard, and a pig (Fig. 4.1). Within the heads of these embryos, on either side of digestive tube or pharynx pockets develop and these extend to the lower face of the skin. During further development, in the shark the pockets open to the outside and increase in size to become the gills. The pockets in the lizard and pig may open for a certain time, but they then recede without turning into gills. These observations lead to the classical conclusion that the pockets in the lizard and pig embryos are the remnants of the adult gills of their very distant fish ancestors.

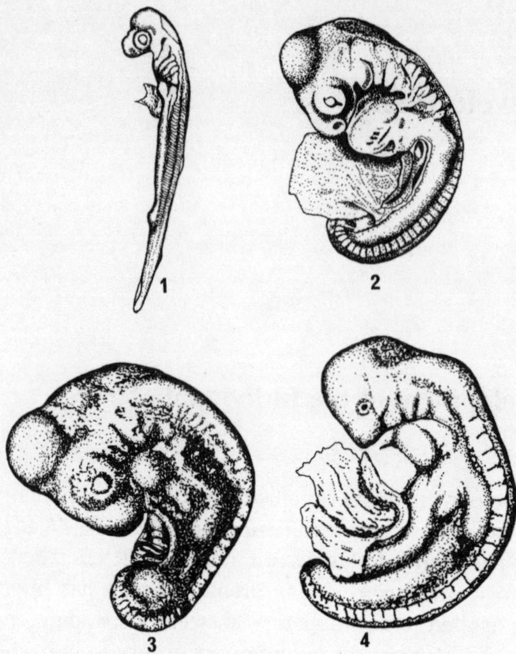

Fig.4.1. Comparison of the embryos of *1* shark; *2* lizard; *3* bird; *4* pig. Each of the embryos exhibits visceral pockets extending into the sides of the posterior part of the head. The similarity shown between the embryonal stages of reptiles, birds, and mammals is quite remarkable. (After Keibel [37].)

From this, and from other observations, the very study of development allows us to state that during evolution, prior to the mammal state, the pig passed through a fish and a reptile stage, a conclusion supported by palaeontological data. Haeckel's law thus possesses a great analytical value and consequently has been widely applied.

However, the case described represents rather an exception. The extensive use of the law meets with so many difficulties, and is subject to so many exceptions, that its significance has decreased and its use has become perilous [3, 4, 5, 6].

To return to the above interpretation, it has to be borne in mind that the so-called branchial pockets of the pig were, at no time during their genesis, really gills and that they thus cannot be considered as an adult stage persisting through the development.

We have to conclude that although the study of phylogeny is a historical, descriptive attempt to study a certain lineage, a series of historical events cannot constitute a mechanism like ontogeny.

Does Development Create Evolutionary History?

We now have to look for a different interpretation and can agree with the statement by Walter Garstang [7] that "ontogeny does not recapitulate phylogeny, but creates it."

An evolutionary history, as defined by adult stages, results from the translation of successive developments which become modified with time. The very mechanisms responsible for these developments are the driving force of evolutionary change.

From this perspective, the supposed "branchial pockets" of the pig may be interpreted quite differently. In all three young vertebrate embryos, the pharynx forms pockets which are now referred to as "visceral pockets"; this is a neutral term, not prejudiced in any way with regard to its ultimate use in the respective adults. From the initial "design" stage common to all vertebrate embryos, the paths of development diverge rather widely. The structures which eventually characterize the shark, the lizard, and the pig are then developed stage by stage. The pockets become gills in the former organism, and disappear in the latter two. As another example, consider the skin folds which turn into fins in the shark and into limbs in the higher vertebrates. Similar interpretations are possible for the heart, the skull, and so on.

Karl von Baer [8], in 1828, correctly interpreted the development of vertebrates and formulated three rules. Firstly: "Features of a less general nature develop from more general ones until the most specialized characteristics have become established."

In the example given above, the functional branchial slits of the shark develop from visceral pockets which are common to all vertebrate embryos.

Secondly: "Instead of passing through the stages of other animals, each embryo of a given species moves further and further away from them." The visceral pockets of the pig embryo do not pass into the branchial slit stage, but rapidly embark on a development of their own.

Thirdly: "Basically the embryo of a higher animal is not comparable to the adult of a lower animal, it only resembles the embryo of the latter." The pig embryo (higher animal) does not resemble a shark (lower animal), but exhibits at its beginning the same structures as the shark embryo.

There is nothing "evolutionary" about these rules. The terms "higher" and "lower" do not imply an ancestor/descendant relationship, but rather a hierarchical concept of a scale of living beings which was widely accepted during von Baer's time, and he himself was not an evolutionist.

Why Has the Synthetic Theory Not Taken the Development Data into Account?

At the end of the 19th century, embryology took a new direction, passing from a descriptive phase to an explicative one (causal or experimental embryology) under the influence of Wilhelm Roux, Hans Driesch, and Laurent Chabry (see A. Dalcq [9].

With the aid of mechanical and chemical methods, researchers tried to influence or modify the course of development, not so much to illustrate certain aspects of particular eggs (for example, sea urchin, mollusc, and frog) but more with the aim of unravelling the mechanisms that are common to all development. In this causal research the question of genetic determinism did not arise [10, 11]. Genetics and embryology advanced independently side by side and the synthetic theory, a concept introduced by geneticists, did not take into account the embryological discoveries.

A combination of the two subjects has come about only during the last few decades when, in experimental embryology, the mechanical and chemical "tools" were complemented by the genetic tool of mutations. From this, developmental genetics became established, a subject which has since already yielded a wealth of results and holds promise for the future definition of evolutionary mechanisms.

What Is the Development of an Organism?

This may be considered as the expression of rules of assembly under the guidance of instructions coded in the genetic material; the very mechanisms of development ensure the execution of these instructions. These mechanisms are here referred to as epigenetic.

The development of an egg, such as that of the frog, represents a succession of events which depend upon one another (Fig. 4.2). At the start, the egg has to be fertilized (cases without fertilization or parthenogenesis are disregarded here) to facilitate the onset of the second event: the division of the egg into numerous cells. This state then, in turn, establishes the conditions necessary for the execution of the next event (gastrulation) during which certain groups or layers of cells acquire some degree of independence from each other and gain some degree of mobility,

Fig.4.2. Development of frog and trout eggs. *1* Start of first egg division; the upper hemisphere is pigmented (*ps* upper or animal pole); the lower hemisphere, rich in vitelline matter, is clear (*pi* lower or vegetal pole). *2* Stage of two cells (blastomeres). *3* Stage of four cells. *4* The egg has now divided into eight cells; those of the upper hemisphere are enriched in cytoplasm and are smaller than those of the lower hemisphere which contains vitelline platelets. *5* Stage of 16 cells. *6* Morulas stage (complete sphere). *7* Vertical section through the blastula stage (*cv* yolk cells). *8* Vertical section through a very young gastrula stage: invagination of cells from the upper hemisphere plus the whole of the lower hemisphere starts at the level of the blastophore (*bl*). *9* Older gastrula stage in section; the blastopore is now circular and plugged by a mass of invaginating yolk cells (*cvi*), making the blastocoele disappear. The cells of the upper hemisphere have become folded in and start to form the intermediate cell sheet (mesoderm) and the notochord. Another cavity has appeared and this will develop into the digestive tract (archenteron). This stage corresponds approximately to that of the trout egg shown in *14*. In the frog egg, the shape of the embryo is not yet developed, whereas it is already perfectly clear in the trout egg. *10* Neurula stage (dorsal view). A continuous (medullar) ridge delineates the medullar or neural plate (*pn*); *bl* marks the position of the blastopore on the posterior side of the future embryo. *11* Transverse section along *AB* through the neurula (*c* notochord; *ci* primary gut; *ec* ectoderm; *ed* endoderm; *ms* mesoderm). *12* The medullar plate is completely depressed. The medullar ridges start to join and then to fuse, enclosing the medullar tube, the future nervous system (brain and nerve cord). *13* Transverse section along *AB* (*ep* ectoderm turned epidermis); the mesoderm has become organized into a mass of muscles in the dorsal part of the embryo which starts to increase in length, already giving an indication of the future shape. *tn* neural tube. *14* Trout egg at the closing stage of the blastopore (compare with *9* for the frog egg).

permitting them to take up clearly defined positions within the egg by combined movement. The establishment of this arrangement is the precondition for the differentiation of the various types of cells into organs, systems, and eventually into the functioning organism.

During all these initial stages of development, the controlling connections between the successive events are rather rigorous, but as development proceeds this rigidity may become relaxed. Whereas in the frog egg, the formation of the sheets is followed by the appearance of the first signs of the nervous system (nerve cord and brain) and then by the formation of the body muscles, these events take place simultaneously in the eggs of the trout and chicken.

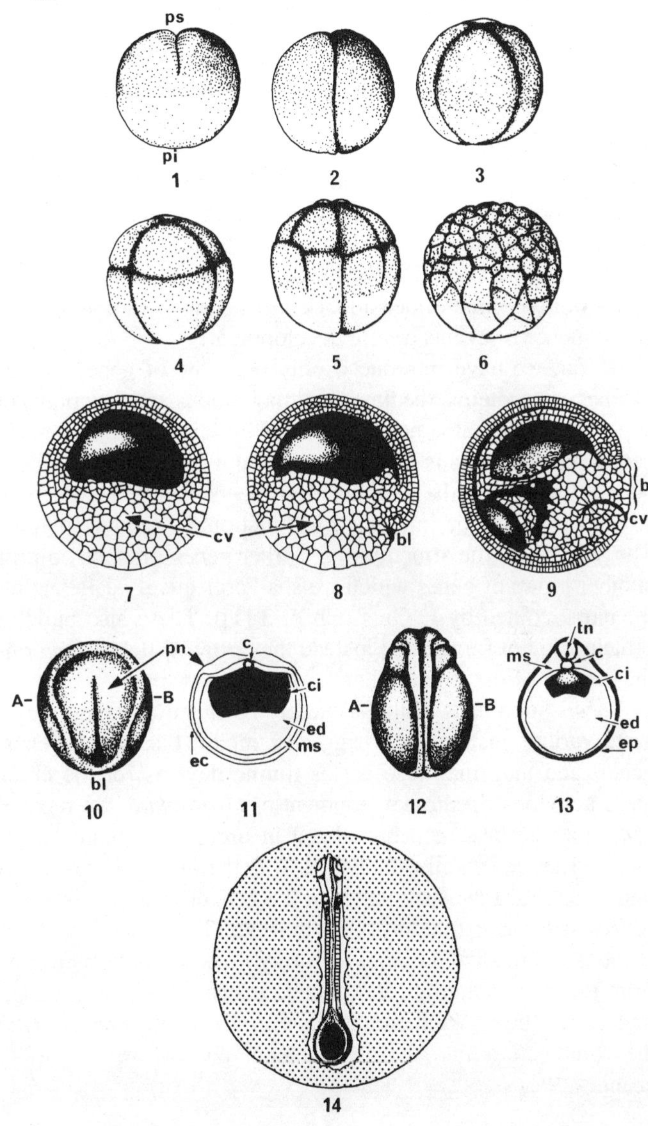

From this comparison of development in different organisms we may conclude that within a chain of embryonal events some, but not all, states may be shifted in time in relation to others which remain chronologically fixed. These potential transpositions are the source of an important evolutionary mechanism, i.e. chronological shift or heterochrony, which may produce changes in the history of certain lineages (Chap. 4, and 10).

We shall see later that there are also other processes in development which may or may not intervene as factors of evolutionary change. This underlines the great importance that is attached, at present, to development as an aid to understanding evolutionary mechanisms.

What Is the Role of the Genes in Development?

The mechanisms by which proteins are synthesized from the information coded in the genes are, by now, sufficiently well understood. But how do the genes intervene beyond this fundamental activity in the execution of a shape or function from a single cell, the egg?

How can we explain how the genes, aligned along the DNA molecule, are able to control the three-dimensional construction within an egg in a strict chronological sequence of events during development?

So far, we have introduced only one type of gene, the structural gene, which synthesizes proteins, the basic building blocks and functional elements of a cell. To supply the material is not the same as building with it. A mass of material is not a cell, a mass of cells is not an organ, and a collection of organs is not an organism. At a certain stage, the various constituents have to be assembled to form coherent structures and to be integrated in functional systems, such as cells or organisms. The activity of the structural or worker genes must be controlled and directed by another group of genes which we shall collectively call "architectural genes", after the name coined by F. Gros (Chap. 3 [1]). These also build up proteins but these proteins do not form cells; instead they provide the instructions necessary to direct the activity of the structural genes in time and space.

These "architects" fulfil a variety of functions: they may detect signals, integrate various instructions, trigger or inhibit (as switch genes) activities of other genes, and they may also act as timing devices for the chronological control of these activities. In the development of *Drosophila,* we recognize that these genes have not just been conceptualized in order to explain the observations, but that their existence is well established. It is the object of development genetics to recognize and localize such genes in the genetic material and to analyse their participation in the construction of an embryo. This science also has to explain how a shape is established, i.e. in which way growth mechanisms operate so that starting from the mere materials and genetic instructions, a leg, wing, organ, or complete organism is built. We therefore have to dissect the chains of events responsible for the construction into their component parts if we are to explain evolutionary changes.

Is Development Controlled Only by Genes?

The genetic material constitutes a programme (software), but is development only controlled by this? Do the instructions describe, in minutest detail, the entire "scope of work", to include all the stages in the formation of an organism?

To adopt such a reductionist approach, which relates everything to the gene, would mean ignoring the way in which genetic instructions are subject to the very mechanism of development, mechanisms which tehmselves are tied to the successive organizational stages that exist between the original egg and the final animal.

We shall illustrate the concept of these so-called epigenic processes in a simple example: A protein molecule consists of an intertwined string of amino acids (Fig. 4.3). The sequence of amino acids which characterize the proteins is determined (coded) by a gene, whereas the mode of intertwining, in part, results from the mutual attraction or repulsion of the amino acids. The three-dimensional arrangements is thus not the direct result of the instruction encoded by the gene, but is the consequence of the initial state of organization created by the gene. This process of spatial arrangement of the protein is described as "epigenetic".

Throughout development, genetic determination and epigenetic mechanisms are continuously active, but in such an intricate way that it is difficult to unravel their interaction. Accepting epigenesis as a fact leads to the conclusion that the initiation of a certain complex developmental process could only by explained by the activation of a limited number of "initiating" genes in a chain of "building activities".

Fig.4.3. Myoglobin molecule. The amino acid chain (primary structure) of this respiratory molecule contained in muscle tissue is shown by the sausage-shaped tube which has been pulled apart in places to illustrate the helically coiled nature of the chain (secondary structure). The "sausage" is twisted (tertiary structure) and encloses the haem group represented by a disc (*h*). This is a chemical entity that extracts oxygen from the blood and eventually makes it available to the cells.

Evidence for the Role of Genes in Development

The so-called "t" mutations in mice mostly occur during the very first steps of development of the egg; these steps are causally connected with each other and are thus indispensable for the continuation of development. Consequently, these mutations are fatal [12] (Fig. 4.4). Is it then necessary to conclude that no genetic modification or early innovation may occur during these initial stages and that, in order to have an evolutionary impact, any such changes would have to take place fairly late?

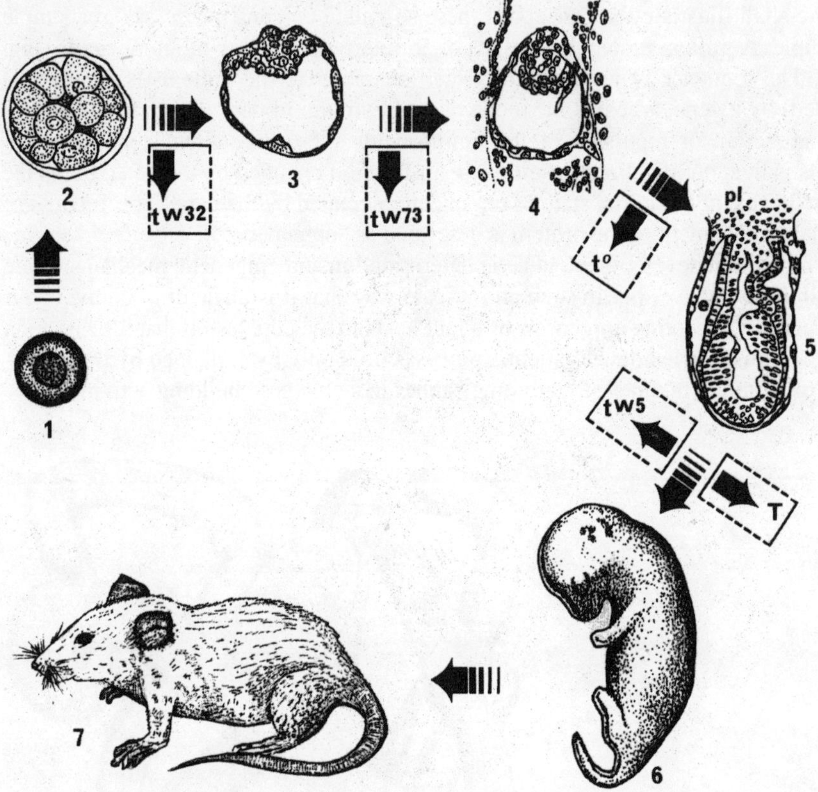

Fig.4.4. The "t" mutations of the mouse. Early stages in the development of a mouse egg and the effects of the "t" mutations. The fertilized egg (*1*) starts to divide (*2*) and forms a vesicle, the blastocyst (*3*). The cell mass will become the embryo while the placenta will develop from the thin wall. *4* The blastocyst has passed through the opening of the Fallopian tube into the uterus itself and has started to settle on the uterus wall where the placenta is formed. *5* More advanced stage showing how the cell sheets of the embryo start to form individually (*pl* placental cone). *6* Very young embryo. Mutation *TW32* stops further development prior to the formation of the blastocyst. *TW73* stops development after implantation of the blastocyst into the uterine wall. *t°* inhibits the separation of the embryonic ectoderm and endoderm. *tW5* prevents growth of the ectoderm. With *T* as single dose T/+ (+ represents the normal gene), development continues to produce a viable animal possessing a short tail. Doubled to T/T, it leads to the degeneration of the embryo. (After Gachelin [12] and the authors.)

On the contrary: the observations made by Rudolpf A. Raff [13] on eggs of various species of sea urchin have shown that innovations may actually take place at any time. In most of these species the egg contains only a small amount of nutritional reserves (vitellus or yolk) and results in a swimming larva, the pluteus (fig. 4.5), which feeds itself and metamorphoses to the adult form. In certain other species, an early innovation leads to the formation of a larger quantitiy of yolk. The embryonic development is then changes radically; the larval stage does not appear and the construction of the ultimate sea urchin begins without metamorphosis.

How Does a Drosophila Form?

Initially, the egg is a small mass of yolk surrounded by a single layer of cells (Appendix 4.1) from which the larva and eventually the adult will develop. The role played by the genes has been the object of many studies, but will be outlined once again here [14, 15, 16].

It is only the maternaly derived genes that initially determine the principal axes of the future larva, back – front –, dorsal – ventral, and left – right, during cell multiplication. Within the cell layer itself, maternal and paternal genes then lead to the formation of grooves which separate the segments that will constitute the body of the larva and of the adult organism (Fig. 4.6).

Subsequently two groups of genes, which are located in the so-called Antennapedia and Bithorax complexes, start to intervene. They direct the particular modelling of each of the body segments. On the second thoracic segment, for example, among other structures a pair of legs and a pair of wings will develop. On the third segment a pair of legs and a pair of halteres or stabilizers, small vibrating bodies used in flight, will form (Appendix 4.1).

A more detailed analysis than that given in this brief sketch reveals the existence of gene interactions which control the activation of groups of structural genes. This confirms the presence of the multiple regulatory genes, mentioned ear-

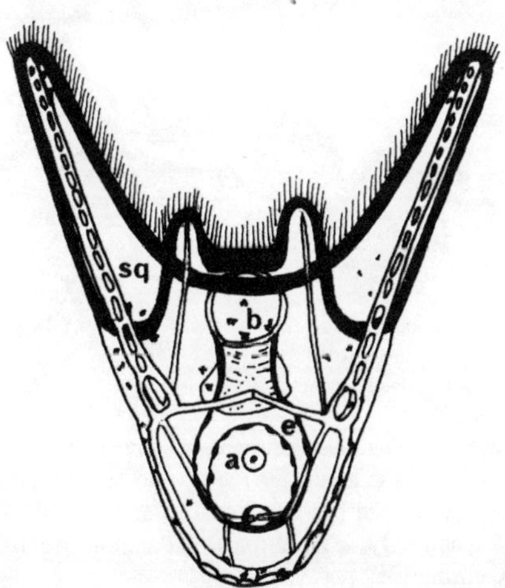

Fig.4.5. Pluteus larva of a sea urchin. *a* Anus; *b* mouth; *e* stomach; *sq* calcareous spines or spicules forming the supporting skeleton of the larva. The ciliate band (*black*) is used for swimmin movements.

lier, and their distribution in functional levels as proposed by Davidson and Britten (see Chap. 3, Fig. 3.4).

Starting with these observations, Robert A. Raff and Thomas C. Kaufmann [17] proposed a model for the sequential evolution of the genetic material which eventually leads *Drosophila* as summarized in a series of body plans (Fig. 4.7). The earliest members of this sequence have been known since the mid-Cambrian, about 530 million years ago, and are represented in recent tropical faunas by *Peripatus*. Its head is made up of three segments and the remainder of the body by a variable number of rather similar segments, each carrying a pair of legs. From this type of organization and from our knowledge of genetic determinism in *Drosophila,* we may reasonably assume that the genetic make-up of these early animals already contained the genes responsible for segmentation of the body.

The millipedes appeared during the upper Silurian, about 400 million years ago. More segments were added to the head which was then made up of six segments that became fused with one another. In *Drosophila,* the genes responsible for this transformation are located in the Antennapedia complex, which thus appeared for the first time during this epoch.

The most primitive wingless insects, such as the silverfish, made their first appearance during the mid-Devonian, some 380 million years ago. Compared with the preceding stages, the subdivision of the body has advanced to a thorax of three segments with true legs and an abdomen of seven segments with rudimentary appendages. This situation leads to the conclusion that new genes have appeared in the Antennapedia complex and that the newly established Bithorax has started to control the reduction of the abdominal appendages.

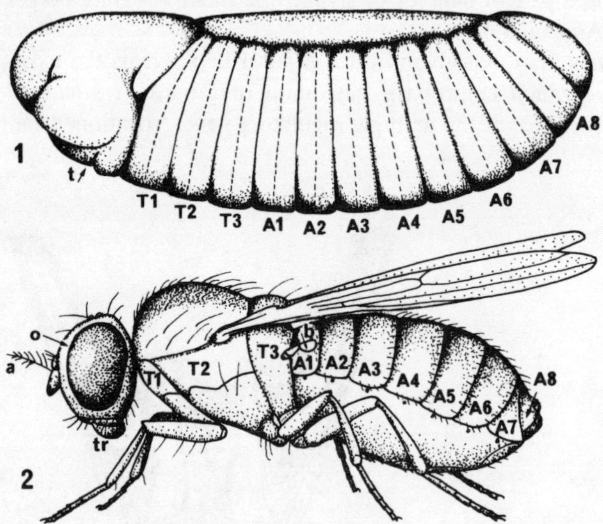

Fig.4.6. The development of *Drosophila. 1* Larva after the formation of the body segments. Those of the head (*left*) are fused to a varying degree. *T1–3* thoracic segments; *A1–8* abdominal segments; the hind portion of the body will form the genital apparatus. *2* Body segmentation of the adult fly. *a* Antenna; *b* balancing organs or halteres; *O* composite eye; *tr* proboscis. (After Gehring [38].)

In the upper Carboniferous (320 million years ago) the first winged insects, comparable to the present-day dragonflies, appeared. Thoracic segments 2 and 3 had each developed a pair of wings. However, the genetic programme for the differentiation of insect wings is not yet known. To encounter insects with two wings, like *Drosophila*, we have to wait until the early Jurassic some 195 million years. The wings of the third thoracic segment were replaced by the stabilizing halteres. As shown by the analysis of *Drosophila*, the transformation of a wing to a haltere results from the activation of other genes in the Bithorax complex.

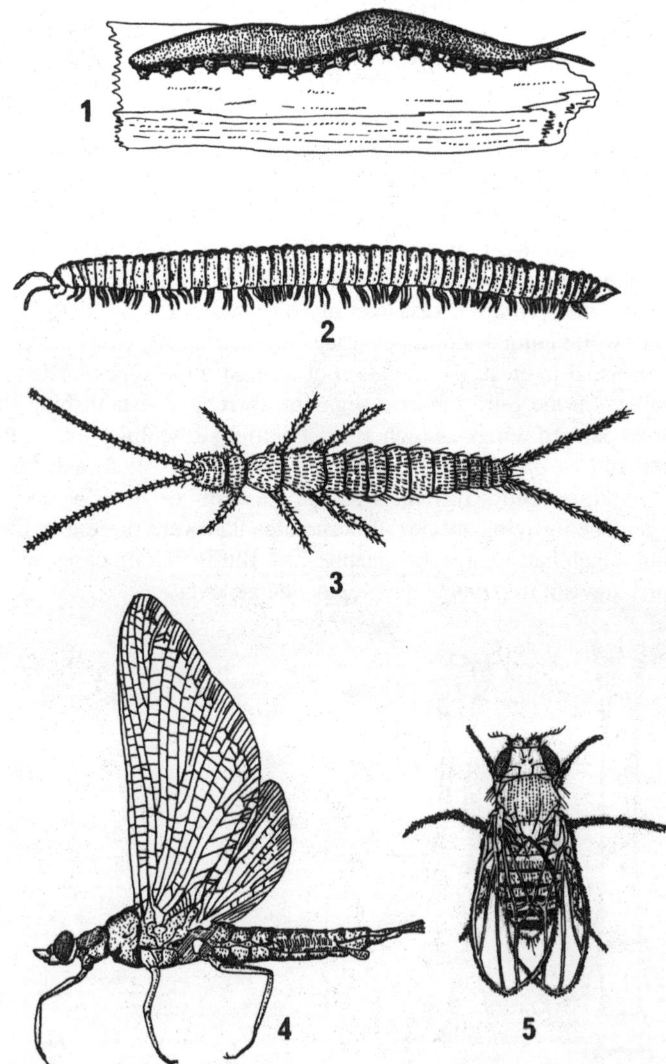

Fig.4.7. Some stages in the evolution of the insect body plan. *1* Onychophore: *Peripatus*. *2* Myriapoda: millipede. *3* Wingless insects: springtail (order Collembola). *4* Primitive winged insect: mayfly (order Ephemeroptera). *5* Dipteran insect: *Drosophila*.

The above interpretation is still rather hypothetical and incomplete, but it could certainly be refined by further studies. Its essential merit lies in the fact that the classical morphological evolutionary history complements the genetic evolution. It is based not on "mutations", postulated to explain the observations, but rather on genetic activities of the present and which, supposedly, must have been developed in the past.

Atavism: The Reappearance of Lost Features

Let us return to the genes which define the nature of each segment of the fly, i.e. that one segment carries the antennae, and another a pair of legs or a pair of wings. These genes may exhibit rather curious properties: the genes responsible for the formation of the halteres, for example may permit the re-appearance of a pair of wings, a structure characteristic of the second thoracic segment. On the head, a mutation may result in an antenna being replaced by a leg. These, and certain other transformations, represent a "derailment" of the development system (appendix 4.1); however, this is not without interest for evolution.

In *Drosophila* the vast majority of mutations result in "monsters", as in the insect world animals do not bear legs on their heads. However, one particular case is of special interest: the mutation of at least three genes leads to the appearance of wings on the third throacic segment, thereby re-establishing the tetrapteric stage (two pairs of wings), which at least partially resembles that of the supposed ancestors of *Drosophila* (Fig. 4.7). Such a mutation is called "atavistic".

Until recently atavism was only a term to describe the spontaneous reappearence in living species, of structures that were present in the ancestral species, but which had long since disappeared. But in certain cases at least, we may interpret atavism in terms of developmental genetics.

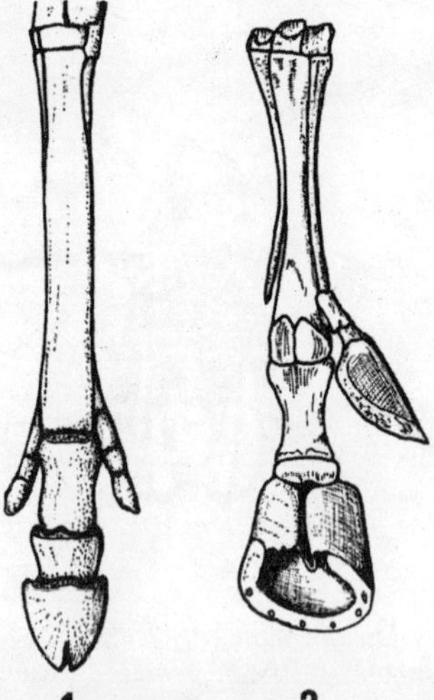

Fig.4.8. An example of atavism. *1* Leg of *Merychippus*. *2* Right foot of the domestic horse with a well-developed lateral digit. (After Wood Mason *in* Rensch [18].)

Examples of atavism are rather numerous. Present-day horses sometimes develop the lateral toes II and IV which were present in their Tertiary ancestors (Fig. 4.8). This was reportedly the case in Alexander the Great's war horse, Bucephalos.

These remarkable cases pose a more general problem: During evolution what has become of these systems (genes or cells) which formed the original structures that have now disappeared? The second pair of wings in *Drosophila* disappeared more than 160 million years ago. The laterial toes of the horse turned into non-functional bony splints less than 10 million years ago, and so on. And now these organs may reappear spontaneously! This shows that the systems responsible for their construction have been maintained intact, but have lain dormant until some genetic event (one or more mutations) reactivated them. How could they be conserved over millions of years?

The system for the formation of wings on the third thoracic segment of *Drosophila* was reused for the halteres; but what is the explanation in the case of the toes of the horse?

In other cases a certain "disorganization" has set in to varying degress. A system, for example controlling the formation of the mouth in birds no longer initiated the formation of teeth after the end of the Cretaceous, some 65 million years ago. However, in experiments, the system still exhibits the ability to form enamel, but not dentine [20]. In contrast to this, rather complex genetic systems have completely disappeared. Despite the numerous cases of reptiles and mammals returning to aquatic life, the gills of their distant fish ancestors have not yet reappeared.

How Does Development Intervene in Evolution?

Chronological Shifts in Development

A number of genes have been observed to determine the chronological sequence of the various stages of development and may be held responsible for the phenomena referred tp as "heterochronies" [6. 21. 22. 23].

A well-known example is the ambystome, a North American salamander (Fig. 4.9). The animal lives on land but lays its eggs in water. The egg forms a swimming larva which metamorphoses and transforms into a terrestrial animal that

Fig.4.9. The aquatic *axolotl (below)* and the terestrial *ambystome (above)*.)After Duméril [19].)

becomes sexually mature and lays eggs so that the cycle starts again. The Mexican axolotl, a sort of large larva of the ambystome, spends all of its life in water and also breeds there. Metamorphosis has thus disappeared from its life cycle. The axolotl may be considered as a stabilized larva, but it is possible to experimentally induce it a metamorphosis to a type of ambystome which moves onto dry land and becomes sexually mature.

In the life cycle of the ambystome, sexual maturity follows metamorphosis. For the axolotl, a chronological shift of these events (heterochrony) causes maturity to set in prior to metamorphosis which, consequently, will no longer occur. The animal thus remains for its entire life in a larval state. This shift is controlled by a so-called temporizing or P-gene [21] (Appendix 4.2).

The importance to evolution of these shifts in development lies in the fact that the respective processes may be initiated by minor genetic actions (a sole gene in the above case) which lead to radical changes in the morphology of the adult form. Furthermore, such transformation is abrupt and does not show any intermediate stages, as would be expected in gradual evolution. It is thus a discontinuous evolution which does not necessarily require natural selection (Chaps. 9 and 10).

How do these temporizing genes work and how do the cells recognize time? These questions remain unanswered.

The Leg of the Toad

Adult organisms compete utilization of the environment; The result is natural selection. Similiar, so-called physiological competition [24, 25] should occur much earlier in life cycles, starting with cell groups that eventually form organs. This has been illustrated by experiments on the development of the limbs in toads [26, 27, 28], in which the number of toes has been reduced (Fig. 4.10).

A reduction in size and number of toes is frequently observed in lineages of higher vertebrates. This might reveal the phenomenon of embryonal competition [29] which could be of importance as an evolutionary process.

If, in contrast to this, the quantity of cell matter is artificially increased, as shown experimentally by Hampe [30] (Fig. 4.11), a certain skeletal element like the fibula which became reduced in birds since the *Archaeopteryx* some 150 million years ago, may reappear.

Growth and Evolution

Growth may be analysed either in the context of its genetic determinism or from its effects on the final form that an organism acquires. It is this latter aspect that will be discussed here.

Let us return to the development of the toad: Up to the end of the gastrula stage, the shape and volume of the egg do not change much. With the start of neurulation and the use of the embryonic reserves, the embryo starts to grow in length and with different regions increasing at different rates to establish the shape of the embryo. These growth processes are controlled by two mechanisms; namely, a multiplication of the number of cells and a lengthening of cells themselves. These mechanisms may operate concurrently with in the same structure.

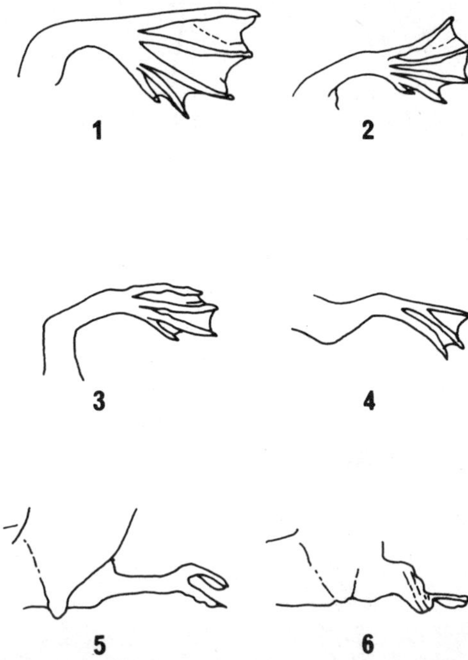

Fig.4.10. Bretscher's experiment. There is a reduction in the normal number of toes in the leg of *Xenopus* after treatment with colchicine, which reduces the number of cartilage-forming cells. *1* Normal leg. *2* Simultaneous reduction in length of all toes. *3* Disappearence of toe I. *4* Disappearance of toes I and II. *5* Disappearance of toes III and IV (?). *6* Preservation of toe IV (?) only. This experiment demonstrates that the anlage of the various toes compete with each other (26) when the quantity of cells required for the formation of the skeleton falls below the normal value. This is comparable to a competition between animals for an insufficient quantitiy of food. In this "struggle", the various toe anlagen possess different capacities. The first toe shows barely any resistance whereas the fourth, the strongest, attracts at the expense of the others the meagre resources remaining after the extended treatment. (After Bretscher and Tschumi [26].)

We can now study the absolute growth of an organism or structure in terms of its size or weight as a function of the individual age (development) or of the geological age (evolutionary history). A consideration of the latter led to the supposed "law" of Cope–Depéret, often termed "Cope's rule", relating to the size increase in certain lineages (Chapt. 12).

A study of the growth of one structure (y) compared with another (x) illustrates the change in proportion with the variation in size of an entire organism or of certain of its components [31, 32] (Figs. 4.12 and 4.13; Appendix 4.3).

In certain cases the analysis of relative growth facilities the explanation of a type of evolution erroneously referred to as "exaggerated growth" (or hypertely), which is supposedly fatal for the animals concerned. The case of the Irish Elk (Fig. 4.14) is one of the products of this misinterpretation.

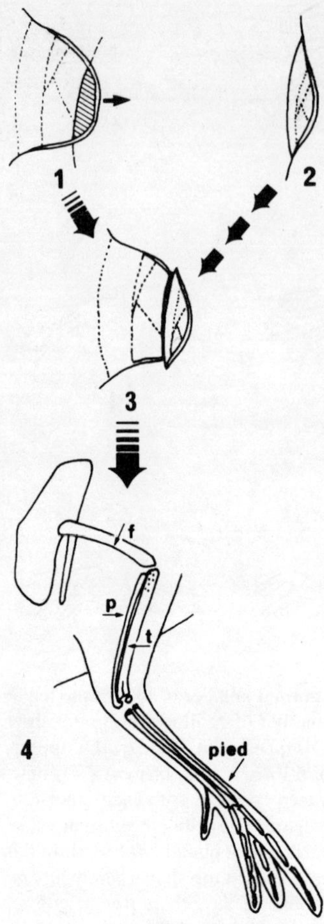

Fig.4.11. Hampé's experiment. *1* In a chicken embryo, the outermost part of the bud of the posterior limb (*to the right of the line*) is cut off and (*2*) replaced by a complete bud removed from a younger stage. *3* From this composite arrangement, (*4*) a posterior limb is formed in which the skeleton exhibits a well-developed fibula (*p*); in a normal limb this bone is present only as a small bony rod. *f* Femur; *t* tibia. (After Hampé [30].)

The adult male has spectacular antlers which are usually interpreted as being so bulky that they would have condemned the species to death. This view is based on the following line of reasoning: For some unknown reason antler growth started to escape any selective control and, with all restraints broken, evolution led along a fatal course to exaggerated growth. Such an interpretation is, however, completely unfounded! What does "exaggerated" signify here? Exaggerated in comparison to what? A study of the relative growth of the antlers in various extant and fossil cervids [33] has shown that when the size of the respective animal increase, the antlers, do not grown in direct porportion, but at a greater rate. The Irish Elk, the largest known cervid, bears antlers that are very much in proportion to its body size. That such an evolution should have escaped selection is an unfounded assumption. However, it cannot be excluded that the size of the Irisk Elk was indeed controlled by a selection which maintained an equilibrium between the advantages of a larger

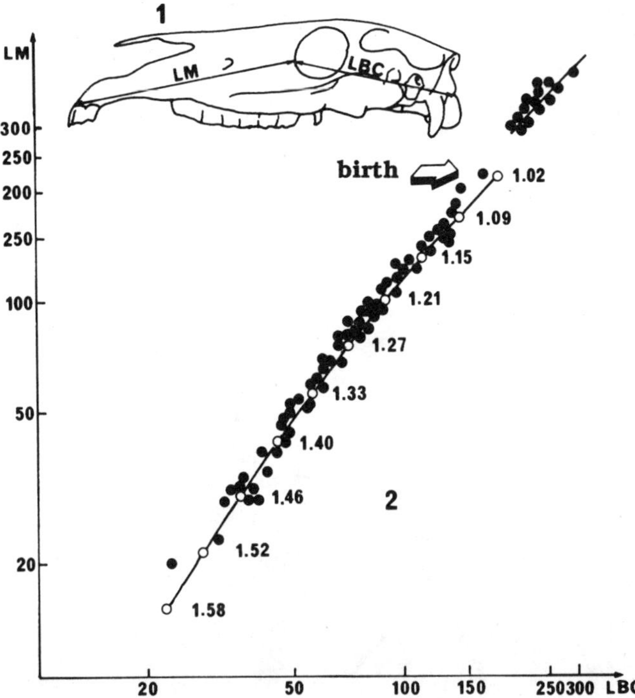

Fig. 4.12. Allometry of horse skulls. *1* Position of skull parameters. *2* Length increase (y) of snout (*LM* in part *1*) vs. length of the brain case (x = *LBC* in part *1*). Equation: y = Ax2 + Bx + C. The data points define a parabola. The numbers to the right of the curve, representing the slope of the tangent, show that the relative growth rate decreases progressiely from positive allometry (1.58) to virtual isometry at the end of the foetal stage (1.02 at birth). The following points represent measurements of the skull during juvenile growth (see also Appendix 4.3). (After Devillers et al. [32].)

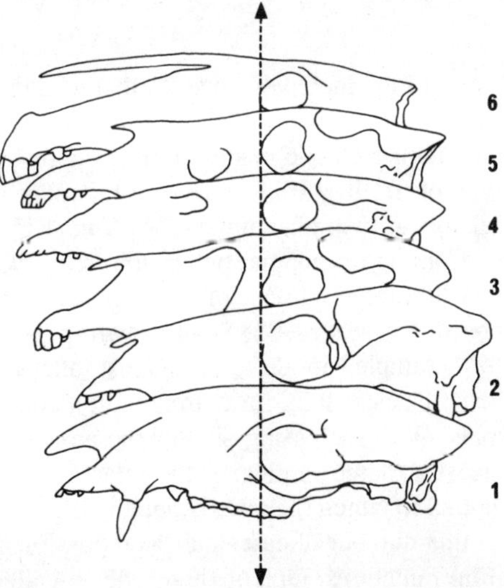

Fig.4.13. Skulls of fossil horses. Lateral view of skulls of some genera of equids aliged at the front edge of the orbit and illustrating the increase in facial length vs. the length of the brain case during phylogeny. *1 Hyracotherium* (Eocene). *2 Mesohippus* (Oligocene). *3 Merychippus* (Miocene). *4 Gripphippus (Miocene). 5 Equus* (Present). *6 Ilippidion* (Pleistocene). (*1* after Gingerich [39]; 2–6 specimens from the American Museum of Natural History (New York), after Devillers et al. [32].)

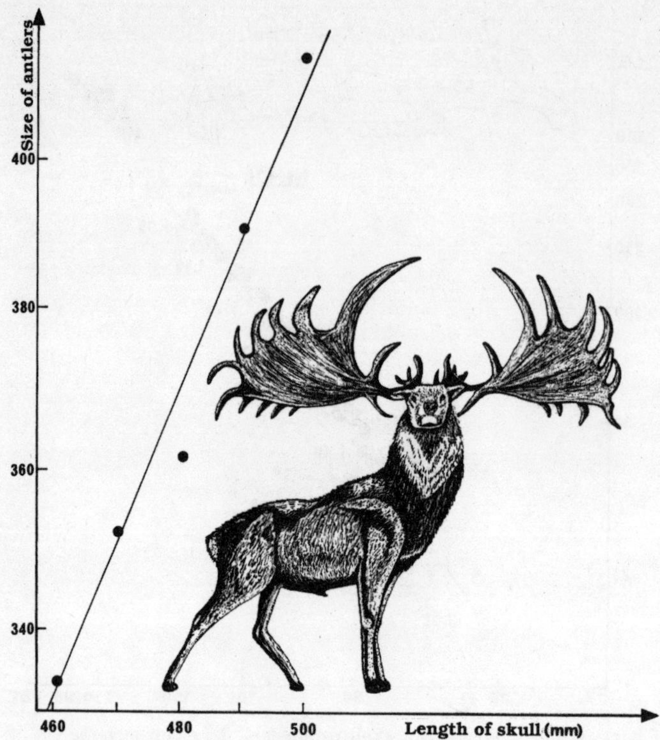

Fig.4.14. The Irish elk. Reconstruction of *Megaloceros* and graph showing the size of antlers vs. length of skull. The positive allometry (Appendix 4.3) is responsible for the enormous development of the antlers concurrently with an increase in skull size (and thus also in body size). (*1* After Knight, a painting in the Am. Mus. Nat. Hist., New York; *2* After Gould [33].)

size and the inconvenience (?) that resulted from a large demand for energy during the annual renewal of this enormous mass of bony matter.

Another classic example of "exaggerated growth" occurs in the titanotheres, a type of Tertiary rhinoceros and this may be correctly reinterpreted in the light of relative growth phenomena [34] (Fig. 4.15).

Contrary to popular belief, the Irish Elk, titanotheres, and other examples, show that in a large number of lineages selection favours increased body size (the supposed "Cope's rule"). Such growth presents certain physical advantages such as, for example, in defence against attack by predators, and also physiological benefits, especially for animal with a constant body temperature such as the mammals. With increasing size of the animal, the ratio of surface area to volume decreases, thereby reducing the relative loss of body heat; this contrasts to a small animal in which the ratio is notably higher.

In a number of cases has been possible to gather data on the genetics of growth from mutations. One of these, the so-called basset ("ancon") mutation in dogs, is

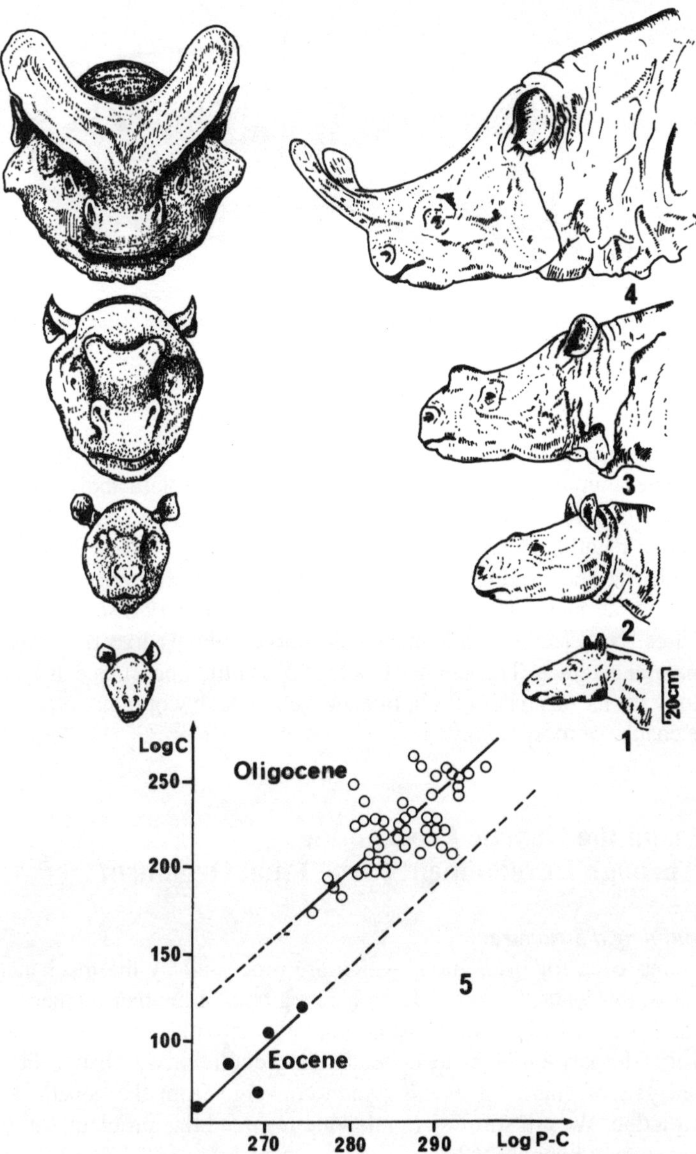

Fig. 4.15. Evolution of the titanotheres. *1 Eotitanops borealis* (lower Eocene), size of a large dog. *2 Manteocerus manteocerus* (mid-Eocene). *3 Protitanotherium emarginatum* (upper Eocene). *4 Brontotherium platyceras* (lower Oligocene), height at shoulde 2.5 m. *5* Length of horn (y axis) vs. length of cranial base (x axis). The lengths, expressed as logarithms, are presented on an arithmetic scale. For the Eocene specimens, the allometric equation is $y = 5.09x - 12.76$, and for the Oligocene ones it is $y = 4.59x - 10.83$. the two coefficients of allometry are thus similar (5.09 vs. 4.59). The evolutionara change is given by the distance between the intersection on the ordinate from the origin of the axis (12.76 vs. 10.83). For the smaller forms, body growth did not continue long enough to allow the formation of horns. As the period of growth was extended, body size increased and horns were able to develop. (*1–4* After Osborne, (in [34]; *5* after McKinney and Schoch [34].)

Fig.4.16. *Teleoceras.* The short-limbed rhinoceros from the Miocene and lower Pliocene of North America. (After Knight)

of special interest. It also occurs amongst sheep where it results in short-legged viable animals [35]. This mutation may also explain the appearance of the rhinoceros *Teleoceras* during the Tertiary; *Teleoceras* was short-legged like the above-mentioned dog (Fig. 4.16). This "mutant", however, was able to survive under natural conditions over a longer period. The importance of this observation lies in the fact that an abrupt change or evolutionary discontinuity may, in certain cases, be selectively advantageous. Large animals like the rhinoceros have few natural enemies. The lenght of its legs is of little importance as the animal does not look for its survival in flight like a gazelle which would not be able to survive such a change of morphology.

From the Genetic Programme Through Development to the Final Organism

Building a Structure

Genes code for instructions which are processed by the mechanisms of development. this leads, for example, to haemoglobin or another pigment, to cartilage, to a leg or group of legs, or to the coordinated activities in the body of a viable organism. The key problem to understanding evolutionary change lies in the detailed analysis of chains of events which converge from the genetic level to the final structure. We are still far from having resolved this problem, but what is the present state of knowledge?

Even though for a molecule like haemoglobin we can follow the stages of its synthesis step-by-step, we still do not unterstand how or what initiates the process, or why certain cells start to develop at a certain time. For a complete structure like a leg, the problem acquires further complexity with the process of differentiation where a well-ordered arrangement shapes various groups of united cells into skeletal, muscular, or nervous elements.

The organs which have been studied in great detail, at least in their early stages, are the limbs of higher vertebrates. From the results of such studies [36] we are ablo to develop the following scenario: Regardless of whether the animal usually walks, swims, runs, or flies, the skeleton of the front appendage shows a common basic design: upper arm, forearm, and hand. This common design suggests some kind of underlying "software" for a higher vertebrate limb. It is modified by a number of processes that direct the running of the programme to produce the correct structure. What do wc know about this software and its directive systems (Fig. 4.17)?

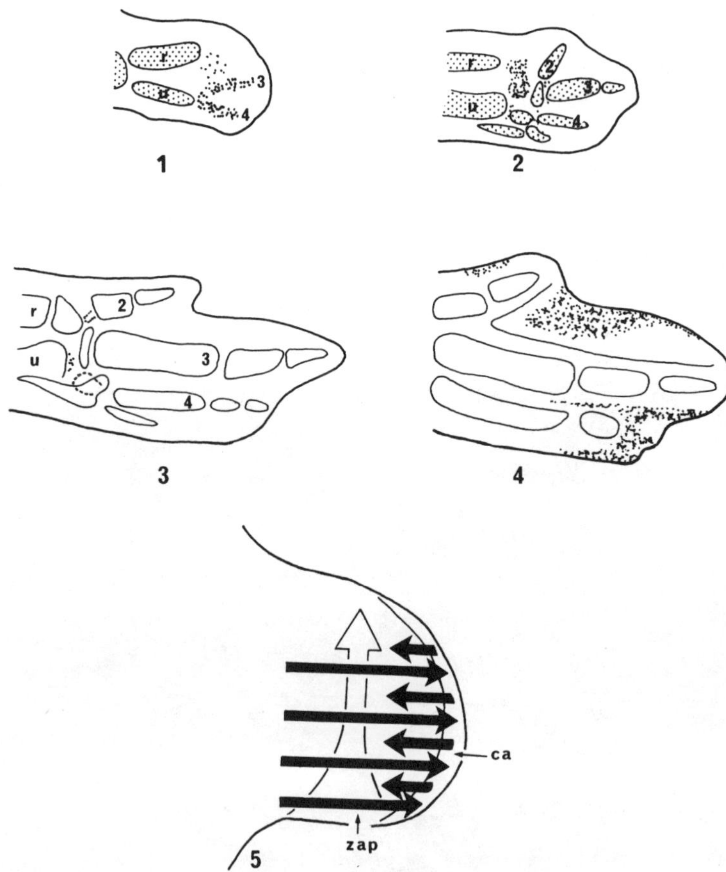

Fig. 4.17. Development of the front limb of the chicken. *1* Establishment of cartilaginous rudiments of the radius (*r*) and ulna (or cubitus) (*u*); cell concentrations outlining the third (*3*) and fourth (*4*) metacarpals. *2* The cartilaginous rudiments of metacarpals *2*, *3*, and *4* are formed., and the carpal elements start to appear. *3* At this same time all skeletal elcments of the hand have become established. Thereafter, various fusions take place, eventually leading to the adult construction. *4* On the rudiment of the limb, zones of cell necroses *(stippled)* start to shape the final limb. *5* Scheme of the Saunders-Zwilling hypothesis for the actions and interactions (*arrows*) of different zones of the limb rudiments on the formation of different skeletal elements. (*arrows pointing to left*) the apical ridge (*ca*) initiates the growth of the skeletal mesenchyme, from which a factor (*arrows pointing to right*) controlling activity and survival of the ridge emerges. The zone of polarizing activity (*zap*) controls the direction of the anterior–posterior axis of the limb. (After Hinchcliffe [36].)

Following the constructions pathways of a complete limb, with its skeleton, muscles, blood vessels, and nerves, poses a problem, the complexity of which still defies all understanding. However, step by step scientific investigation finds its way through this maze and our present knowledge, scant as it is, nevertheless permits us to draw some interesting conclusions about evolutionary mechanisms. From what we already now know about the actions of genes and of activating zones, we are able to conclude that the formation of a structure offers various possibilities for intervention by factors that modify the "design" and thus can establish evolutionary change. This provides further proof of the importance of embryological studies, whether they be decriptive, causal, or genetic, in our understanding of the role of determinism in evolution.

Chapter 5

Adaptation: The Driving Force or a Result of Evolution?

What Is Adaptation?

Lobsters and dolphins are adapted to aquatic life whereas the green woodpecker and the fly are adapted to aerial life. The way in which these organisms are constructed allows them to execute certain functions within certain environments. This is a first, simple definition of adaptation which describes:

The evolutionary process of adjustment of an organism to its environment. This historical aspect will be discussed in subsequent chapters.

The result of this process – the state of adaptation – or the form of an organism at a certain point in its history. This functional–structural aspect will be dealt with here.

By the very fact of their past and present existence, organisms are, or were, adapted to the prevailing conditions on Earth. Life itself is an adaptation. However, this statement is too general to be of use to us. We need to consider the particular features of a certain environment and ask ourselves how the general adaptations to life in water, air, etc. were established.

Often, a consideration of adaptation is restricted to a single structure: for example, the legs of the green woedpecker for clinging to a tree trunk, or the legs of the praying mantis for catching prey. In this rather restricted approach of considering only one component, the entire organism as such is completely ignored.

Whether it be general or specific, a particular adaptation may be executed in several ways, either by organisms with different body plans (phyla) or by organisms with the same body plan (classes, orders, etc.).

A "question" about the environment will not lead to a single "answer" only, but to several answers which form part of the adaptive strategy.

Between similarly adapted animals there may be comparable features which result from the presence of common functions: The wings of the fly and of the green woodpecker fulfil the same function (analogous organs) but are constructed differently (non-homologous organs). These common features are accompanied by differing, specific features that result from particular adaptations: compare the legs of the fly with the legs of the green woodpecker, or the proboscis of the fly with bird's beak.

An analysis of different types of adaptation poses the following questions: Will all the components of an organism participate equally in its adaptation, as postulated by the panadaptionist approach, according to which everything leads to adaptation? Is it not evident that some structures are indispensable for a function whereas others are less important or even neutral? To what degree will an organism execute a response to the demands of its environment? Is it possible to say that an adaptation is optimal, or "perfect", or is it just a compromise between different, possibly contradictory, constraints which result from the construction of the organism in question and the conditions of the environment? And lastly, is adaptation the driving force of evolution?

How Can Different Organisms Adapt to the Same Conditions of Life?

Let us consider as an example adaptation to aquatic life: Ponds, rivers, and the sea are accessible to representatives of virtually all current phyla. The requirements for aquatic life may be satisfied through quite different anatomical constructions and by patterns which are a function of the degree of complexity of the organism in question. We shall restrict ourselves to considering respiration and, more particularly, the supply of oxygen.

In whichever type of organism we consider, the problems of gas exchange in either an aquatic or aerial environment are controlled by the physicochemical process of diffusion of gases through cell membranes into the cytoplasm.

For sponges and hydras (Fig. 1.4), which are exclusively aquatic, and for the somewhat more complex aquatic or terrestrial platyhelminths, diffusion across the body surface will satisfy the respiratory requirements. Thus one, cannot talk here about an anatomical–physiological adaptation to respiration in water.

Annelids also utilize diffusion acros the epidermis, but they increase the efficiency by having a blood circulation and pigments which transport oxygen deeper into the body. This is no longer a particular adaptation to aquatic life: The necessity of a circulating system results from the size and complexity of these animals. However, the marine worm *Arenicola* (Fig. 5.1) has developed a very simple aquatic respiratory apparatus in the form of branchial tufts which are rich in blood vessels.

With the still more complex molluscs and crutaceans, the gills, an aquatic respiratory organ, became established. Both the blood circulation system and the gills are found to be nearly unchanged in fish (Fig. 5.2), where they attain a remarkable degree of efficiency for the extraction of oxygen [1].

The cetaceans have returned to an aquatic life, adapting to it in a number of ways as is shown, for example, by the skeleton, caudal fin, etc. However, they still breathe air but have certain modifications that permit them to dive for one hour or more.

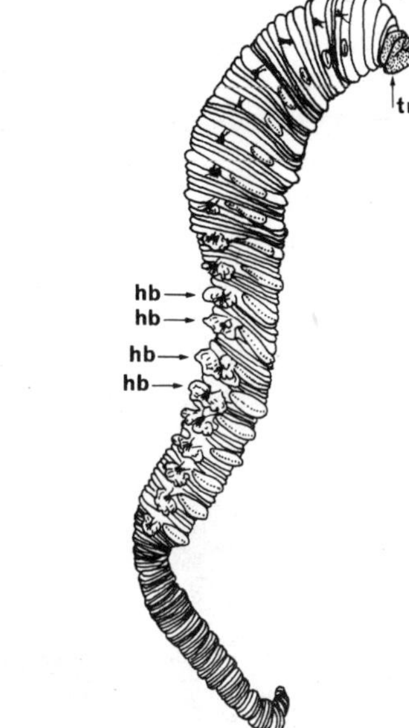

Fig.5.1. *Arenicola. hb* Gill tufts; *tr* proboscis (retracted).

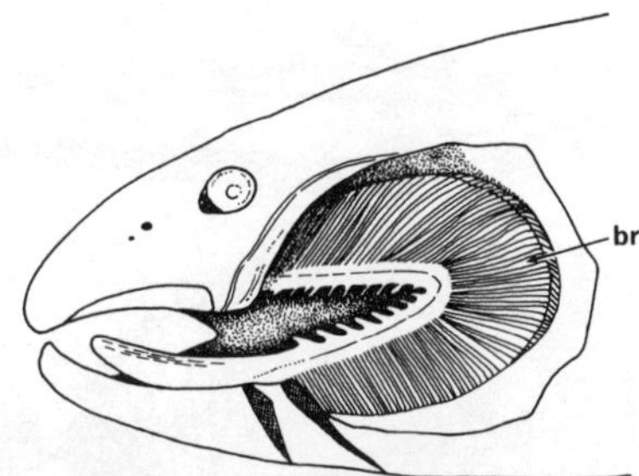

Fig. 5.2. Gill apparatus of a fish. The operculum has been removed to show the gill filaments (*br*).

Remarkable Adaptations

Successful aquation organisms, such as some crabs [2] and fish [3], have adapted to breathing air – an astonishing and, at the same time, paradoxical adaptation.

In fish, the gills play an important role in this adaptation. They are made up of lamellar rows of gill filaments which, with a small volume, present a considerable surface area for gaseous exchange. They are indispensable for maximal extraction of oxygen from water which contains only about 0.7 mg/l of this gas at 15 °C. In air, the gills can no longer function because they do not have water pressure to hold them up. The various layers collapse onto each other and the respiratory surface is dramatically reduced. Also, the epithelium covering the gills dries out, thereby becoming impermeable to oxygen.

So, we may ask the question: What has caused certain fish, especially those from freshwater, to acquire aerial respiration?

An explanation which is frequently advanced is that selection may have favoured those mutations which allowed fish to escape from unfavourable conditions, resulting, for example, from desiccation of their environment or from a low oxygen content in the water due to higher temperatures. Both in the past and at present this occurs especially in tropical areas.

This idea is plausible but fails to explain all the aspects: Although most fish with the ability for aerial respiration do live in tropical areas, some are also found in temperate waters or in the sea where these dangers do not exist. Furthermore, only some tropical fish have developed organs for aerial respiration and, remarkably, they belong to different groups and have developed their respiratory systems from different organs (Fig. 5.3).

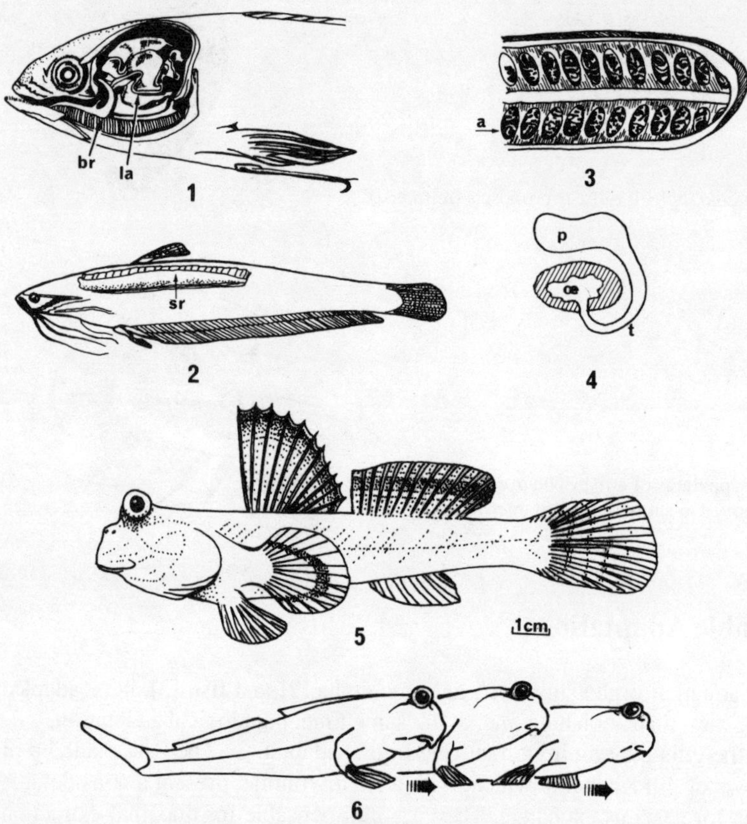

Fig.5.3. Some adaptations to aerial respiration in fish. *1 Anabas* (climbing perch) from freshwaters of tropical Asia; operculum removed to show the branchial chamber containing normally developed gills (*br*) and the labyrinth (*la*). *2 Saccobranchus* from freshwaters of tropical Asia; body wall removed to expose the respiratory sack (*sr*). *3 Neoceratodus*, an Australian freshwater lungfish; part of the lung opened to show subdivision into alveoli (*a*). *4* Transverse section illustrating the oesophagus (*oe*) from which the trachea (*t*) connects to the lung (*p*). (In Johansen [3].) *5 Periopthalmus*. *6 Periopthalmus* moves on land by jumping, using its pectoral fins as feet. (From authors in Johansen [3].)

These, and other arrangements which must have developed in the past, represent a kind of "tinkering" [4], which was certainly efficient but was without evolutionary impact. Only the lungs became established during the Palaeozoic as organs adapted to terrestrial life. They are found in two groups of "fish", namely the fossil and present Dipnoi (lungfish) and thc fossil crossopterygians, the possible ancestors of terrestrial vertebrates.

"Quality" of Adaptations

The body plan of a fish enables it to live in water. But, to what degree do the different organs participate in the establishment of this adaptation?

It can be shown that the respiratory system, the locomotive apparatus and, in a less obvious way, the kidneys, are the essential components of this adaptation. Scales are also a characteristic feature of the fish, but they are not vital for the adaptation as they may be reduced, or even missing completely. Their shape and ornamentation are individual characteristics that are without clearly evident adaptive value. The digestive tract is adapted according to the particular feeding habits.

We could repeat this analysis of the contribution of different body components to the adaptation of innumerable types of body plan and still arrive at the same conclusion: At any level, a living system is a functional complex, the various components (structures, systems) of which participate to varying degrees in the overall adaptation and in various subadaptations. Whilst some components are indispensable, others participate only to a certain degree, and still others do not play any recognizable role.

However, care has to be taken when stating that a certain structure, frequently of a vestigial nature, did not play a role, as the role may only be revealed on closer study (Fig. 5.4).

One should, however, not succumb to the other extreme, that is, the concept of panadaptationism, according to which each organ has a positive influence on the degree of adaptation and thereby increases the chances of survival of its bearer. This view, subscribed to by certain biologists, has been vigorously rejected by S. J. Gould and Richard Lewontin [5]. An organism is an entity, an integrated sum of structures and functions, and it is this complex entity that is adapted as a whole. For the benefit of specific investigations we may, with strong reservations, isolate certain systems or organs and ask ourselves about their role in the functioning of the whole. We can do this for the gills of fish or for the locomotive systems but, at the same time, we should recognize that there are limits to the degree of investigation, such as in the case of scales. Do the intermuscular bones, which are characteristic of the majority of fish, participate in adaptation or not, or are they just a peculiarity of the process of ossification in this group, a peculiartiy, the significance of which, if there is any, has escaped our notice?

In conclusion: The mere fact that an organism is adapted does not imply that every feature contributes to its adaptation.

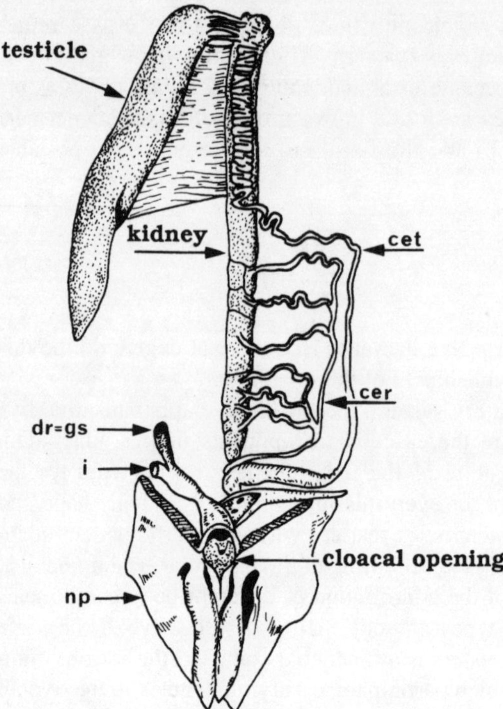

Fig.5.4. An example of a "vestigial" organ in the dogfish. male urogenital tract in ventral view (right half): *cet* evacuating testicule duct; *cer* excreting kidney duct; *dr* rectal diverticle or salt glands (*gs*); *i* terminal portion of the intestines; *np* pelvic fins

A question that is frequently raised is whether "a certain animal is well or poorly adapted to a certain environment". There is no answer to this, as there is no method of measuring the absolute value of an adaptation, despite some ultra-theoretical attempts to do so, however, these are not applicable to the case. We may however, compare in a competitive situation the adaptive values of two species in the same environment. Thus, in temperate climates, the adaptive value of the grey rat is found to be greater than that of the black rat, as the former has managed to replace the latter in a large part of its habitat, although without eliminating it completely. In warmer climates the situation is reversed and the black rat is better adapted.

There is not an absolute value of adaptation for an animal. It can only be defined in comparison with other animals and as a function of the environment conditions (Fig. 5.5).

The adaptive value changes during the history of different groups without us being able to recognize any reasongs for this change. In every case in the history of the organic world there is a fundamental sequence of appearance, expansion, and extinction. This sequence may be onserved at the level of species, genera, orders, classes, and eventually in the different phyla.

The hippurites, reef-building lamellibranch molluscs known as rudists (Fig. 5.6), were perfectly adapted to the conditions of certain seas, such as the ancient Mediterranean, during the upper Cretaceous. They proliferated during this period

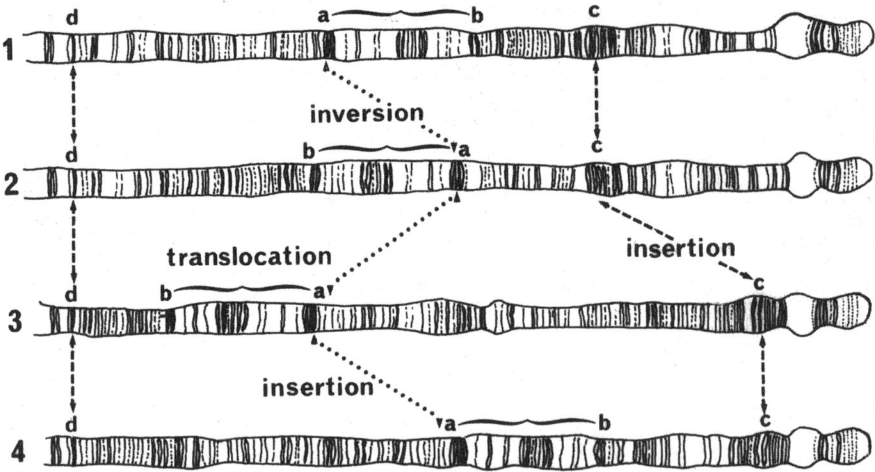

Fig.5.5. Chromosomal mutations of *Drosophila.* Part of the third chromosome in a cell from the salivary glands of larvae of four different varieties of the western American *Drosophila (Drosophila pseudo-obscura).* Chromosomes from such cells are particularly large, exhibiting well-determined sequences of dark and white, wide or narrow bands which have been arbitrarily subdivided into a number of sections. In 1–4 the pattern *d* is always in the same position in relation to the righ-hand end of the chromosome. In *2*, the sequence *ab* is inverted while *c* remains fixed. In *3,* the inversions of *ab* is maintained, although it is shifted, or translocated, towards *d.* Bands have been inserted between *ab* and *c*, shifting *c* towards the end of the chromosome. In *4,* an insertion has shifted *ab* towards the end, shortening the distance between *ab* and *c*, possibly by deletion of certain portions. The four varieties of *Drosophila* are externally indistinguishable from one another. Dobzhansky et al. [6] have shown that the various chromosomal translocations confer on their bearers different adaptive abilities for environmental factors both of a permanent nature, such as altitude, and of a cyclic nature, such as temperature and humidity. (After Dobzhansky and Boesiger [6].)

Fig. 5.6. *Hippurites*. A late Cretaceous reef-builing lamellibranch mollusc.

but disappeared completely at the end of the Mesozoic. But why? The usual answer is that they could no longer adapt to the new conditions. This is merely a truism. We can only state that they did not survive into the Tertiary, but are unable to offer an explanation for the fact.

The "dinosaur" design persisted for over 100 million years during the Mesozoic until it became extinct, together with the hippurites, ammonites, etc. Dozens of explanations have been advanced for this observation, but none is really convincing [7]. Was it an inability to adapt? Suffice it to say that they did not succeed because they could not continue to live. As far as the fossilized remains are concerned, we are virtually unable to recognize the factors which were most critical to their adaptation.

Some recent examples demonstrate which adaptation, above all others, is more likely to render its bearer vulnerable to change in the environment and thus expose it to the threat of extinction. The giant panda, a carnivore (Fig. 5.7), lives and feeds only in the bamboo forests of China. Any change leading to the disappearance of the bamboo will also result in the extinction of a large number of pandas, which are dependent on this particular forage. This happened some years ago when the bamboos eaten by the pandas all flowered at the same time (a very rare event) and then wilted depriving the animals of their basic food stuff.

Fig.5.7. The giant panda. A "carnivore" from China which feeds exclusively on certain species of bamboo, and is thereby strictly tied to its environment.

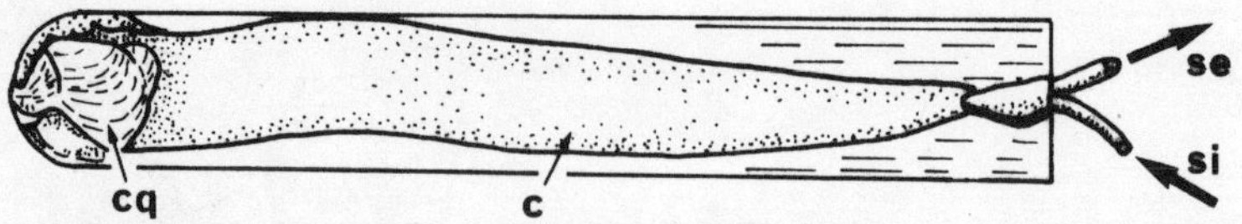

Fig.5.8. Teredo, the shipworm. A bivalve mollusc living in holes which it cuts into wood with the aid of its reduced shell (*cq*). It respires by sucking in water through a tube, the inhalent siphon (*si*), and then expelling it through the exhalent siphon (*se*); *c* body of animal.

The grey rat, with its highly eclectic taste for habitat and food, is better armed than the panda. If it survives a climatic change, or a natural or artificially induced catastrophe, then it will end up the better for having discovered a new living area.

Rabbits were introduced in to Australia in the 19th century and have since proliferated. From the start they were able to adapt to conditions previously unknown to them. At the same time, they caused some smaller marsupials to disappear, although not the kangaroo. There are numerous examples of organisms which reveal a capacity for adaptation in new situations to which circumstances have exposed them.

The adaptations, whether very general or very specialized, do not usually cause their bearers to lose the characteristic features of their original body plan. A whale is still very much a mammal, and the teredo (Fig. 5.8), which lives by boring into wood in the sea, can still be clearly recognized as a bivalve mollusc like the oyster.

Only a parasitic life style, which leads to the exclusive dependence of an organism on its host, may in some cases result in a complete loss of the original identiy of the organism. The liver fluke (Fig. 5.9), a sheep parasite, is clearly still a plathyhelminth ("flatworm"), albeit slightly modified. *Taenia,* of the same phylum, has been much more modified, having lost its digestive tract and relying entirely on its

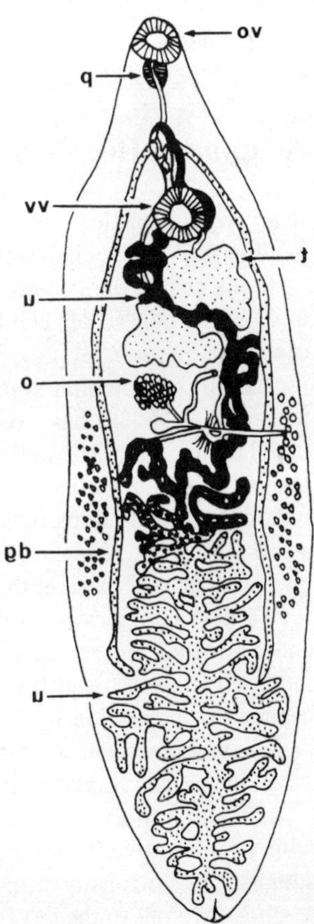

Fig.5.9. *Dicrocoelium,* a platyhelminth liver parasite in sheep. *dg* Digestive tract; *o* ovary; *p* pharynx; *u* uterus; *t* testide; *vo* oral sucker; *vv* ventral sucker.

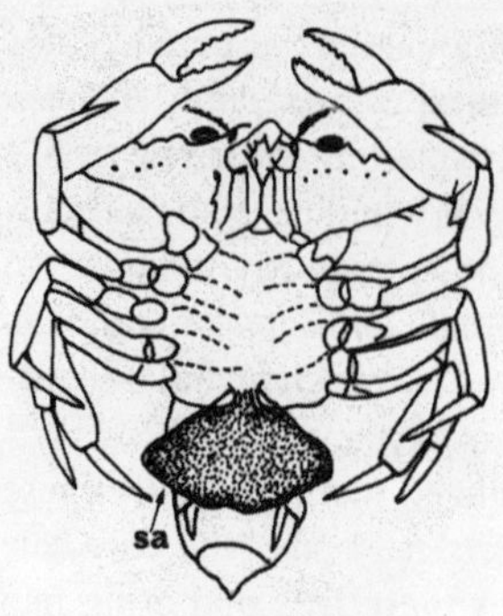

Fig.5.10. Ventral view of a crab with parasitic sacculina (*sa*).

host for nourishment. But to what organism is the sacculina (Fig. 5.10), a sack filled with sexual products hanging from the abdomen of certain crabs, related?

Nothing in the adult morphology permits us to unravel the nature of this undefined sack. Only the passage of the life cycle through a free larval stage identifies sacculina as a cirripede crustacean like the barnacle, *Balanus*. In such cases adaptation reaches its highest degree of specialization (actually "perfection") through degeneration.

Adaptation Does What it Can

Is an orgqanism flexible to the extent that it can always give a better response to the demands of the environment?

In his *Origin of Species,* Darwin wrote: "Natural selection operates only for and through the benefit of each being... All its corporal or mental attitudes tend to proceed to perfection."

Although perfection is referred to several more times in his work, it is regrettable that it should have been used at all in a book of such calibre. The term is based on an unrealistic anthropomorphic approach to what might actually be adaptation. Perfection would reach a stage which could not be surpassed. But how would we be able to recognize this state? Perfection cannot be a scientific concept as, in reality, any adaptation must be a compromise between the constraints of an environment and those of the body plan of the organism.

We are dealing with evolution when we go from one functioning organism to another which possesses components that may sometimes have been radically modified in design and function. The evolutionary process works like a tinker, to use an apt expression by F. Jacob [4], but a tinker puts together pieces here and there, modifies them, or makes new ones, and from this bric-a-brac, a concrete mixer or a small vehicle will result.

In contrast to this, at the start of any evolutionary process there is an organism with its own genetic programme, mechanisms of development, structural constraints, and with its own history, all of which control its adaptive responses. The result is then put to the test by selection, a system of constraints which are external to the organism. The new form emerging from an evolutionary process will rarely differ significantly from its ancestor, with the changes being channelled in a certain direction. Will this new form then be able to occupy ecological niches that are far removed from the original source niche?

In its genetic make-up, *Drosophila* has thc ability to losc its wings. However, in Hawaii, an area of intense speciation of this fly, no species has developed this rather special adaptation which would be favoured, or at least tolerated, by the island situation of numerous insects. This adaptive compromise is not possible for *Drosophila,* although it is carried out by other insects.

Gazelles, antelopes, etc. escape their through flight. Such rapid sustained running on limbs constructed for pendular motion requires two conditions, namely, a lenghtening of the limbs and a reduction in their weight. Taking into account the resistance of bony matter, the weight of the animal, and the constraints acting on the limbs during galloping, the dimensions of a bone of ideal lightness and solidity may be calculated. However, the actual bones are two to three times wider in section, to enable the animal to cope with sudden hazards while running over irregular terrain [8]. Thus adaptation is clearly compromise between the often contradictory constraints of constrution on the one hand and conditions of use on the other hand.

Cetaceans in their adaptation to swimming have managed to modify the shape of their body and their limbs, and have even developed a caudal fin which is moved by particular muscles. However, they were not able to change their respiratory system. The genetic constraints of the "cetacean" life style and body plan have not permitted a total adaptation to aquatic life, as is seen in the fish. This recognition of a limitation of adaptive capacity leads us to conclude that there is a general problem for any evolutionary change: namely, the constraints of expression (Chap. 6).

Adaptation – The Driving Force of Evolution

This principle implies that adaptation is the starting point of speciation, the basic process of evolution. Nobody would deny that adaptation is a condition of life. However, that adaptation is the sole mechanism for speciation is debatable.

In the gradualist theory of evolution, in which one species is continuously adjusting itself to environmental conditions and gradually transforms into another species, adaptation is the prime factor.

But what is its role in the "founder population" model (Chap. 7), in its derivaties, such as "punctuated equilibrium" (Chap. 11), or in the so-called bottleneck model (Chap. 9)?

Is the formation of an isolate from a parental population the result of a tentative exploration of a hitherto unoccupied environment (founder population, punctuated equilibrium), or is it caused by a random event (bottleneck) that is responsible for a reduction in the size of a population? In these cases, the starting mechanism for speciation is not immediate adaptation. We do not know under which genetic or environmental influences a founder population will extend exploratory "pseudopodia". Adaptation to the new environmental conditions is a secondary process, but it will "decide" whether the establishment of a new population is successful.

The effect of a mutation (Chap. 3) does not show any relationship between "necessity" and the viability of an organism. The genetic change itself does not represent an adaptation, whereas its evolutionary success does depend on the ability of the mutant to adapt to its new life.

"Early innovations" or "heterochronies" (Chap. 9), however, fall into the category of initiating the process of adaptation–specification. The "fundamental principle" should therefore be reformulated: "Depending on the particular case, adaptation plays the role of an initiating (primary) or modulating (secondary) mechanism."

Chapter 6

A First Resumé: Organisms, Selection, and Adaptation

Genetics and developmental genetics have given us such a wealth of new insight that, at the end of this century, the synthetic theory can no longer be maintained in the strict "orthodox" sense in which it was started.

The new synthesis to be developed should incorporate the data and new concepts that have been gleaned from the increasing number of other sciences that touch on evolution. To give an exhaustive presentation of this new information would go far beyond the scope of this work, and here we shall single out only some of the findings, notably, the constraints on evolutionary change, the concept of functional levels in the genetic material, the economical use of genetic information, the connection between the genetic programme and development, the various levels of integration, the execution of shapes and structures, and continuity versus discontinuity in evolution.

Not Everything Is Possible in Evolution, Even with Time

Throughout the preceding chapters we have encountered the concept of constraints which limit the possibilities for expressing the variability, and thereby the execution, of evolutionary change.

The concept of constraints is fairly old. Cuvier [1] had already expressed it at the beginning of the 19th century in his "law of correlation of shapes". According to this law, compulsory relationships exist between the structures of an animal; these relationships will permit only a limited number of combinations, to the exclusion of others: A carnivorous tooth controls the construction of the entire animal.

Darwin's natural selection is one expression of these constraints, acting on the animal externally and limiting the number of viable possibilities. These constraints affect only the final organism, but some modern authors have outlined others constraints which act at the level of the genetic instructions and which are active throughout development. This broadened concept of the intervention of constraints leads us to a radical revision of our understanding of evolutionary change.

The possible variations in the genetic programme of a species over time and space are innumerable and the life of a species will be a continuous lottery. The genetic material exposes to chance a multitude of designs which the necessary selective pressure will accept or refuse. Evolution would thus be the *Chance and necessity* proposed by J. Monod [2].

Chance does play a role (see Chapt. 12) but, despite the proposals of certain misinformed authors [3, 4], it is not the sole factor responsible. No sensible evolutionist will ascribe the order of the organic world purely to chance. Natural selection intervenes as some kind of "anti-chance". But is this sufficient?

Is it conceivable that of the innumberable designs exposed to selective pressure, a sufficiently large number will succeed to ensure the survival of the species and allow it to evolve? This might be the case for bacteria, which have an enormous reproductive potential, but is certainly not so for the elephants, which flourished and diversified over some 50 million years of the Tertiary. Their evolution, like that of virtually all other animals, has to rely on other mechanisms. We shall not invoke any mysterious "internal force" which directs the entire lineage to a distant target, the process of orthogenesis proposed by some authors. Rather we shall consider constraints which are a reality. Intervening at every level they act in restricting (narrowing) the range of possible changes (not everything is possible). Construction is channelled so that the organism is basically stable in time and space. The genetic lottery will only lead to modifications within certain limits. For a species to venture outside its usual framework (Chapt. 7) is a rare event for which, at present, we do not know the reason.

An organism, as we envisage it under this new concept, is not an entity passively directed towards its evolutionary target by omnipresent and overpowering natural selection. Due to the constraints inherent to their construction, the individuals of a species will expose only a very limited number of possible designs to the external constraints of selection; thus the lottery is "biased". Along with Gould and Lewontin [5] we may say that by the very mode of its constructions, every organism controls its evolutionary destiny to some extent. Selection also plays a role, but no longer such an exclusive one.

Omnipresent chance or unrelenting determinism? Between these two extremes there is space for a concept according to which the advance of evolution is the result of the continuous interaction of random phenomena, "anti-chance", or determinism.

Let us look briefly at the origin of some of these constraints (Chap. 3, 4, 5).

Connections between the functional levels in the genetic programme link the genes together, thereby limiting their (theoretical) capacity of expression to a larger extent than in the old linear genetic programme.

In a developing egg, the necessary coordination between biochemical syntheses, cellular activities, and interactions between cells, layers, tissues, organs, and systems, in other words the continuous interplay between genetics and epigenetics, will only allow the establishment of certain highly channelled pathways of development. This stabilization of development will permit the organism to soften the impact of modifications by the genetic instructions, usually without final grave repercussions: a frog's egg will always produce a frog, even it is a monstrous one. So far, embryologists and geneticists have not succeeded in changing the development of one species into that of another one. And furthermore, if development was not somehow stabilized, there would be not recognizable species.

The intervention of these early constraints results in the domain of possibilities (the "game of possibilities" of F. Jacob [6] and impossibilities. That a certain body plan is not known from the past or present does not necessarily mean that it has not arisen in the past and been rejected by selection, or that this body plan might possibly be realized in the future. However, the combination could also be impossible to realize.

These considerations lead us to the problem of the existence of forms that are intermediate between two types of body plan. Intermediate forms between two species or two types of body plan, and the transition from one type to another is referred to as obligatory gradual evolution. That gradual evolution is a real mode of evolution may be shown by a number of examples. However, it does not entirely preclude the possiblity of abrupt steps from one form to another through evolutionary discontinuities without intermediate stages. The mechanism of development open the way for such an evolutionary possibility. the switch genes, delineated in studies on *Drosophila,* place differentiation in front of an alternative, either this way or that but without the possibility of connecting from one to the other.

This concept of alternative choices was extended by George F. Oster and Pere Alberch [7] to all levels of the organic world, from the development individual to the separation of evolutionary lineages. They introduced the idea of evolutionary bifurcations which underlines the concept of discontinuities interrupting evolution. The authors proposed, as a theoretical example, the differentiation of the skin in higher vertebrates. One can imagine that from an ancestral dermo-epidermal papilla, a bifurcation of development directed evolution along one of two pathways: either the papilla extended outward during growth and turned into a scale or a feather, or it turned inward to form a cuticular gland or a hair. Between these two pathways there is absolute discontinuity: a scale can never become be hair.

The organic world is neither in chaos nor is it a perfectly continuous system. The various forms may be grouped according to their body plans (phyla, classes, etc.) and are separated from each other by pronounced gaps.

Are these gaps necessary (essential), i.e. tied to the very patterns of evolution and to the occurence of bifurcations during its history, or are they contingent and provisional, resulting from the incomplete state of our knowledge? There is evidence to support both cases.

From a stock of primitive vertebrates a bifurcation took place during the Palaeozoic, resulting in the evolution of the Synapsida (which led to the mammals) and the Sauropsida (from which the "reptiles" – an artificial class – and the birds were eventually derived). In modern nature "reptiles" and mammals are clearly distinct. However, from palaeontological data we know that in the past there was some continuity and the separation was thus contingent. In contrast to this, the gap between mammals and birds is substantial.

Thus, evolution may be discontinuous and the presence of intermediate forms is not a prerequisite. Furthermore, what about those numerous supposed "intermediate forms" that are widely used to stress problems with the evolutionary theory or even to deny it? They are essentially products of the human imagnation. No one fossil can be the intermediate form that is hoped for, as there will always be one feature missing or another feature that is too far evolved, or not far enough. Nothing ever conforms to the norm. But what is normal? We may define it from the following postulates: any intermediate stages could and should have existed, a postulate which disregards the constraints of possibilities and impossibilities; and all parts of an organism evolve at the same speed, so that halfway through their evolution the representatives of a lineage like the mammals should have advanced all their structures to the same state in, for example, the level of transformation.

This, however, is an unrealistically rigid concept which suggest a coordinated evolution and is in contradiction to any study of the palaeontological record. Throughout their evolutionary transformation, the components of an organism possess a certain freedom which allows them to evolve at rather different rates. From this observation, Gavin R. de Beer coined the phrase the "mosaic law of evolution" [8].

Consequently, we must profoundly revise our view of the pathways in evolution.

Genetic Programming Works Economically

The genetic programme does not encode all the details for the construction of a structure or an organism to the minutest detail. Its instructions establish the rules of emplacement of the molecular complexes within the space of the egg and its course of development. However, there are other events, like the association of various proteins with one another or with other molecules, which do not appear to have any immediate genetic determinism. These processes rely on chemical affinities and are the so-called epigenetic mechanisms. Furthermore, new properties which result from the state of organization of the egg during the various stages of construction of the embryo manifest themselves during development.

The synthesis of a certain protein is coded for by a gene or part of a gene. A chain of reactions, like the metabolic cycle, activates a number of proteins or enzymes and, consequently, a number of genes, although the same enzyme may participate in several cycles.

The sole action of one or more temporizing genes can, by way of chronological shift, initiate events which have a profound evolutionary impact, and possibly generate discontinuities.

One would have to concede that with the hierarchical functional organization of genetic material, the establishment of new regulatory systems may play a much more important role than the mere multiplication of structural genes, which number about 3000 in the coli bacterium and may number about 100000 in the more complex humans.

Mutations that affect one or a small number of genes may lead to spectacular physical changes such as the transformation of the stabilizing halteres into wings, or to smaller changes like that of the colour of the eye. Pleiotropism will further diversify the impact of any one mutation.

In conclusion, we can say that not every detail or step in the construction of an embryo will require the intervention of one or more genes to effect the execution of this detail. We may state that the genetic material operates economically with, in most cases, one instruction leading to the initiation of a whole chain of processes.

The extent of an evolutionary change is by no means a direct funtion of the number of genes involved. The genetic programme is able to execute it all, and there is no need to postulate, along with Stephen M. Stanley [9], that there is a different type of genetics for macroevolution from that for microevolution. There is but one type of genetics!

Genetic Programmes and Development

Carl H. Waddington insisted repeatedly in his publications that it is necessary to take development into account when considering the delineation of an evolutionary process, and in his research he established a first basis for developmental genetics [10]. Goldschmidt also devoted himself to this new reform of the synthetic theory, but his arguments did not find wide acceptance due to the dubious manner of their expression [11].

Waddington illustrated his concept of a complex functional programme of genetics/development through an "epigenetic landscape" which outlined a number of fundamental points. The topography of the model is of a Jurassic" type: Parallel ridges are separated by confluent and divergent valleys and are cut by necks connecting the valleys. A developing egg is visualized as a ball rolling down the slope of a well-defined system of valleys, necks, and bifurcations which it will follow during organogenesis.

This topography, however, is variable: the valleys will deepen or become filled, the necks will become lower or deeper, and new ones will appear or old ones disappear. The course of the ball will therefore be modified, so that it embarks on different routes. This change in topography is caused by the genetic programme which, like a shunt, controls the course of the ball, keeping it to the usual pathways and leading it eventually to the final result, an individual that conforms with the rest of the species. This is the process of canalization. However, mutations may lead to accidents along the way, thereby modifying the destination.

Although today this conceptual model must be considered as somewhat simplistic, its pioneering merit has to be conceded: it indicates the permanent obligatory connection between the genetic programme and development, two systems which are all too often studied separately. These form a functional unit, described as the "epigenotype" by Waddington. This apt term, which unfortunately has not been retained, should be reinstated.

Hierarchy in the Construction of Organisms

The processes taking place in living beings at the microscopic level do not differ at all from those observed in the physics and chemistry of inert systems. It is only at the macroscopic level that the particular properties will appear which result from the constraints imposed by the need of the organism to reproduce and to adapt to certain conditions. We are thus dealing with an interpretation of processes common to living beings and dead matter as a function of the particular state conferred on them by their origin and their destiny (Jacob [12]).

This quotation introduces to our view of evolutionary mechanisms, the operational concept of the "level of integration".

In reductionist logic, evolutionary change may only result from variations in gene frequency. This impoverished logic ignores the fact that an organism cannot be described solely on the basis of its genetic material, and that it is built up progressively by the elaboration of increasingly more complex structural assemblies, which make up the so-called levels of integration.

The 20 amino acids which make up the proteins of living organisms each have their own structure and characteristic properties. Their arrangement in chains gives them the higher level of complexity of the protein molecule and this spatial arrangement results from the properties of each of the constituents. Among other properties the protein exhibits, for example, enzymatic actitivy which as such are not present at the preceding level of the amino acids. At this higher level of complexity in the protein molecule one or more original properties have emerged which, although dependent on the preceding steps, were not directly deducible from them.

Proteins and nucleic acids, together with other molecules, arrange themselves in coherent systems on a still more complex level of integration; namely, that of the cell with its new properties, of which auto-reproduction is probably the most important.

Once fertilized, the egg cell starts to form an association of cells, which go through the various stages of development such as the morula, blastula, etc. The latter state prepares for the formation of the gastrula in which the cellular layers become mobilized, thereby facilitating organogenesis. We are thus observing the execution of increasingly more complex stages in the spatial arrangement of the cells with their subdivision to perform various tasks.

What is the implication of these levels of integration to our problem? It is the progressive emergence of properties which, as they become steadily interlocked, cooperate to produce the final levels of integration in the development from an embryo to an adult.

Among its properties each level includes certain constraints which channel the pathways of development, thereby assuring their auto-regulation and permanence. At the same time, we may envisage that each of these levels of integration can offer various possibilities for producing changes which, due to the very coherence of each system, will affect the organization and properties to a varying degree. Development could thus repress, as well as enhance, genetic modifications.

This view of organic construction goes beyond the mere variation in gene frequencies, in as much as it assigns the importance for development to the mechanisms of evolutionary change. Unfortunately, however, it has not yet been possible to change the pathway of development of one species of that of another. All we have obtained so far are more or less monstrous mutants which still conform to the framework of the species concerned.

To understand development will be one of the essential objectives of future research and should result in the creation of experimental evolution.

If development is a hierarchical process, can we then extend the hierarchical concept to evolution as a whole, as Gould [13] has proposed? According to Gould, evolution is a hierarchical chain of events which, although different from one an other, may also be complementary, and act on at least three decisive levels:

- variation (we have already examined this);
- specification (the formation of a species) which cannot in every case be related only to the adaptive substitution of genes. However, this could reveal, on the one hand, a different type of genetics, an option that would be debatable from our point of view, or on the other hand, it would not necessarily be the consequence of an adaptation but would rather precede it, as in the punctuation model (Chap. 11); and finally
- macroevolution, a process which is rather different from that of gradual transformation or microevolution. From our point of view, there is no justification for distinguishing between two different evolutionary mechanisms. There should only be one which is fundamentally different from classical microevolution.

How Does a Leg Form? A Classical Problem of Evolution

The analysis of certain scenarios of organogenesis has been pushed quite far for *Drosophila*. There are genes responsible for the fixation of the body axes and the delimitation of the various segments, for modelling the characteristic structures of one or another segment, for directing the course of development (binary choices), and so on. The question then arises, how do all these activities become involved in the formation of, for example, a leg with its articulated sections, each covered by a carapace that has a characteristic pattern, with muscles of diverse functions, with nerve fibres, etc.? In short, how does a complex assembly become functionally integrated in a complete organism?

In the formation of the chicken leg the action of various genes has been unravelled and the phenomena of progressive determination of the parts of the limb have been analyzed. They probably intervene by marking the position of cells through the diffusion of certain substances called morphogenes. Although our knowledge of developmental genetics advances in *Drosophila* and the chicken, we are still unable to answer the seemingly simple question: How do form and function become established?

If we should ever be able to outline the series of interlocked mechanisms (a black box) that direct genetic instructions to one structure or an other, then we will have achieved a major breakthrough in our understanding of the mechanisms of evolution.

Continuity or Discontinuity in Evolution

According to reductionist reasoning, a chance for survival will only be accorded to those minute modifications which, without altering greatly the state of adaptation of their bearer, will permit eventual re-equilibration of the adaptation. As we have seen, there are mutations which have a radical effect on organization, such as a shortening of the legs in certain sheep, or a reduction in the wings to vestigial organs in *Drosophila,* and which have been analysed genetically in detail. Another radical mutation is the complete loss of wings in some other Dipterans and other insects, and even in birds, like the Pacific island rails or the "wingless" mutation in the chicken, etc.

Such a structural event unbalances the biology of the mutant. A functioning wing is built to certain proportions and once this design has been changed the animal can no longer fly. It is thus an "all-or-nothing" situation with no intermediate stages possible. The aerial organism becomes confined to the ground.

Will every imbalance of biology be an evolutionary catastrophe? The answert has to come from the field of ecology. If, by chance, the mutant encountered an appropriate ecological niche, it may be able to survive. This is proven by living wingless insects and birds and by the respective fossil examples from the Mesozoic and Tertiary. What are the percentages of hits and misses? We do not know, but from the very existence of such radical changes we can conclude that discontinuities are a possible way of evolution.

Part III
The Message of Fossils from the Palaeontological Record

Chapter 7

The Formation of Species

The Street, an Observation Point for Variation

Let us sit on the terrace of a street cafe and watch the crowds pass by. The first basic unit we observe is the individual: tall, small, fat, long-faced, round-faced – the diversity is large. Among these individuals there will be a few with particular features, such as colour of skin or type of hair. which differ from the vast majority of local individuals, thereby indicating that they originated from another part of the world. We shall install a video camera at this site and will carry out the same observations simultaneously in Paris, Dakar, and Tokyo. When the films are then brought together and shown in the same room, the observers will not fail to note that the individuals of one city have certain dominant characteristic which set them apart from the individuals of the other cities. there is thus a group of individuals who have similar features and constitute a local population. However, despite the particular features characterizing the local population, the various individuals will show a much larger number of common features. Furthermore, they may mix, as shown by couples where the two individuals are from different populations; the children accompanying such couples are proof of their fecundity. The large number of shared common features, and especially the fecundity of these combinations, show that the various individuals belong to the same species.

When observing the fauna in larger regions on our planet, we note the presence of groups of species: biological communities. Individuals may thus be grouped together into three large categories; namely, local populations, species, and biological communities.

What Is a Population or a Species?

Specialists refer to a population as individuals who occupy a local territory and are able to interbreed at random, i.e. “a group of individuals in a situation which allows each couple the same probability of mating and producing descendants” [1]. Local populations capable of interbreeding belong to the same species. This concept appears to be quite evident for man but, frequently, it is rather difficult to define, for example in the case of dogs which, despite their great diversity, still belong to a single species.

The same concept of a species as is presently accepted, has appeared several times throughout the course of development of philosophy and science.

The first observation of the concept of a species is found in Plato's *Allegorics* in which he pronounced that the world is made up of a limited number of permanent, real, and universal "types". Under this concept, organisms appear as true reproductions of these essential types and because of this K. Popper [2] referred to this philosophy as essentialism. This strictly typological species concept, based on the criterion of similarity, profoundly influenced philosophers and natural scientists in the 18th and 19th centuries, as well as unfortunately also influencing some modern naturalists. In the typological approach a species is a rigid and true reproduction of an ancestral type. In modern genetics this would be referred to as "cloning", the process by which groups of identical cells are produced at will from an original source. Such a concept of immutable types is entirely incompatible with our understanding of evolution. It forms the basis of a static vision of the world, corresponding to the fixism favoured especially by Linnaeus and Cuvier.

Negating the existence of such types, William of Occam and his mediaeval school of philosophy developed the so-called nominalistic concept of species. This is distinguished from the preceding concept by the fact that it recognizes the existence of only individuals. Species are thought to be merely human inventions without any objective reality. Lamarck also subscribed to this concept for some time.

A third concept, unanimously accepted today, is the "biological" species concept. It was proposed by Buffon who was the first to introduce the idea that breeding between individuals characterizes a species. This concept found its true expression only during the elaboration of the synthetic theory in the 1940s by E. Mayr.

A biological species consists of populations of similar individuals that form a genetic unit with a large collection, or pool, of genes. It is a reproductive unit in which, theoretically all individuals of the community may contribute to the procreation of descendants. As an ecological unit it occupies a larger or smaller array of biotopes. "Species are natural groups of populations capable of interbreeding, but reproductively isolated from other similar groups." This species concept passes from the mere individual to the population and is compatible with evolution as it integrates the essential aspect of the relation between individuals and their variability.

The biological species concept encompasses an important aspect of biology, namely, the mutual recognition of sexual partners belonging to the same species. This leads to an important secondary consequence, i.e. that the individuals of one species are not able to interbreed with those of another species. Biologists refer to this as reproductive isolation, and it serves to separate the different species from one another.

For a number of scientists the criterion of reproductive isolation appears to be insufficient and, as a result, some have proposed that it should be the concept of "specific recognition of sexual partners". This concept allows us to understand how the individuals of one species recognize each other by various signals and, therefore how the species define themselves. These systems of recognition of partners, operating from generation to generation, define the temporal unity of lineages through reproduction within the species over their entire life spans.

Are Two Species Easily Distinguished Between?

Yes and no! Between the horse and the donkey there are no major problems, as it is effectively the criterion of interbreeding which may be put to the test. However, the distinction between apparently different species is not always so easily defined. If it is reproductive isolation which constitutes the essential distinguished criterion between two species, we have to try and identify by which distinctive features this isolation and the discontinuity accompanying it are defined.

In nature there are some species which appear to be identical in morphology, although they differ in other features, and thus their reproductive isolation is ensured. In these cases we are talking about sibling species. We shall now analyse two examples, one in mice and the other in *Drosophila*.

The Domestic Mouse and the Field Mouse

Mice are an exceptional model for evolution as, next to man, they are the animals best known from nature, for the simple reason that they are also used in laboratory experiments.

In the south of France there are two types of mice: those with a long tail which live in human settlements and those with a much shorter tail, which live in bush country, on dry limestone plateaus, and in open fields [3]. Is the length of the tail, cited by zoologists, of any significance? Does it reflect a genetic difference, a difference between two distinct species?

A detailed analysis of the genetic variability by electrophoretic techniques has shown that the two types of mice do indeed differ in their genetic constitution, thereby indicating that they are reproductively isolated. This statement is supported by field observations showing that they can coexist withouth interbreeding. The length of the tail is thus a valid criterion for distinguishing between the two species. Further studies have revealed that the two species may also be dinstinguished by certain finer details, namely, in the shape of their teeth.

The Sibling Drosophilas of Central and South America

Experience has shown that by detailed morphological studies it is sometimes possible to distinguish between two species on the basis of some minor details that reflect a discontinuity in the species. This statement applies, for instance, to two sibling species of *Drosophila* in Central and South America (Appendix 7.1). They are referred to as siblings because we are not able to immediately distinguish between them in nature. However, they will infallibly recognize each other, during their mating periods.

In this case reproductive isolation sets in at the behavioural level during the mating display, where there is a distinctive signal of recognition. Reproductive isolation also exists at the ecological level because the two species exhibit different temperature tolerances. Furthermore, they show some differences in the morphology of the copulatory organs.

The above examples illustrate that divergences between species may occur at any point in the hierarchical organizational levels of the living world. For example, these may be at the level of genetic constitution (gene mutations), morphology (distinct shapes of certain organs), behaviour (differences in mating displays, aggressiveness), and of ecology (different ecological preferences) [4]. The changes contributing to reproductive isolation are frequently rather variable.

What is the Purpose of Reproductive Isolation?

We now see that reproductive isolation is a factor that permits two species to be distinguised on the basis of sexual reproduction. This isolation is the consequence of the divergence of specific lineages which originated from a common ancestor, but it is a secondary effect of the divergence.

In addition to being an expression of the extent of the divergence itself, reproductive isolation is also indispensable for the separation of species. What would actually happen if there was no such isolation? Let us imagine that some populations have been separated geographically for a long time. Each of them will then develop certain modifications. If these populations came into contact again and that there was a lack of reproductive isolation, the individuals would once again interbreed and the ensuing genetic mixing would very rapidly lead to rather uniform populations. This could also happen to subspecies, as is illustrated by certain Japanese mice. If this were always the case, then there would be no new species. Reproductive isolation is thus of prime importance for evolution, as it permits the multiplication of species and lineages in a restricted territory, and at the same time, ensures the integrity of each species.

When populations with a common ancestor actually come into contact again in the same territory, reproductive isolation allows the descendants to live together independently. At the same time, it precludes the recombination of the two genetic programmes and thereby their mixing.

However, when two populations resulting from the same ancestor do not live in the same territory, the extent of their eventual reproductive isolation is of no importance to the two species as they have no real possibility of meeting and interbreeding with each other anyhow. In this case reproductive isolation is only a "potential" mechanism in case the future history of the two lineages should bring them together in the same territory.

In Mice, Proliferation Outweighs Aggressiveness

Reproductive isolation can also intervene in the processes which precede or follow actual mating between two individuals. With *Drosophila,* we have seen that isolation occurs during the mating display and thus is prior to mating. The possibility of interbreeding is thereby excluded; otherwise it would destroy the integrity of the new species.

To return to the domestic mouse and the field mouse, we can now understand better how reproductive isolation is put into effect. Curiously enough, these two types of mice are able to interbreed if crossed in the laboratory. However, of the resulting hybrids, only the females are fertile whereas all the males are sterile. The hybrid females may then be crossed again with males of one of the two parental species. These experiments allow an understanding of the factors responsible for sterility and allow us to locate the position of the respective genes within the chromosomes. Although these mice are able to interbreed in the laboratory, they will not do so in nature, despite the fact that they may coexist in the same habitat. Why should this be?

When we put one male of each species into the same cage, we will always obtain the same result: a fight will start and the field mouse will win over the domestic mouse, because the field mouse is much more aggressive.

Another experiment, carried out in an enclosure with a simulated natural environment, allows us to further advance in our understanding of this phenomenon. If we put several families of each of the two species into this enclosure, we soon observe that it will become filled with domestic mice. This appears paradoxical in view of the dominance of the field mouse. We note that the domestic mouse, rather than being dominated, will proliferate rapidly as it is able to stand higher densities of individuals. It thus adopts a hierarchical social life, whereas the field mouse, although usually dominant due to its greater aggressiveness, is not able to sustain such promiscuity. Its life style is that of one male living with several females in a larger territory, and is, to some degree, a solitary life. In mice we therefore see that organized proliferation is more efficient than individual aggressiveness.

Variation Within Species

Among the variations within species, we may distinguish variation between individuals from variation between groups or populations.

Variation Between Individuals

This is indeed important and with the exception of true identical twins, there are no two identical individuals within a sexual species. Some variation is linked to the age of the individual (moulting, metamorphosis, etc.), i.e. the different developmental phases, whereas other variation is of a seasonal nature – consider the different winter and summer coats and plumage of mammals and birds respectively.

Other variation appears to be tied to the specific habitat, i.e. the morphology of an organism is controlled by the particular conditions of the local environment. Furthermore, there is the so-called ecophenotypic variation which is reversible, depending on the particular ecological situation. Continental molluscs are especially subject to such variation [5–6], their shells become larger and more elongated as temperature decreases (see Chap. 9).

Variation Between Groups

This is even larger than variation between individuals and is based upon geographically controlled variation within a species. The area of distribution of a species is frequently subdivided into different territories with different ecological conditions. These lead to local modification of populations, which we refer to as subspecies. They may be clearly recognized within a species, although they are still able to interbreed with one another. A good example of this is presented by the famous domestic mouse, which has already been mentioned [3].

The Domestic Mouse. It is referred to as commensal as it is directly linked with man and his habitat, and its world distribution closely follows that of man. Mice are found on ships, trains, trucks, aeroplanes, etc.; indeed, virtually everywhere. These mice *(Mus musculus)* have been subdivided in Eurasia into four subspecies (Fig. 7.1), the genetic divergence of which was determined with the aid of electrophoretic methods. The long-tailed domestic mouse already mentioned *(Mus musculus domesticus)* occurs throughout western Europe, Africa, Arabia, and Australia. Throughout the northern part of Eurasia, from Norway to Kamchatka, we find another subspecies *(Mus musculus musculus)*. South of the mountain ranges from the Caucasus to the Himalayas we find to the west, from Iran to India, *Mus musculus bactrianus,* which is succeeded from Burma to Japan by *Mus musculus castaneus*. These four forms are true subspecies as within each of their regions of contact there is a more-or-less wide zone of hydribization which proves that they are still able to interbreed with each other.

The Japanese mice represent an interesting case of what happens when two subspecies that have been isolated over a long period come into contact again. Japan was colonized by man and mice in two waves. The first, originating from Korea or Southeast Asia, brought the subspecies *castaneus,* whereas the second, coming from northern China, was accompanied by the subspecies *musculus*. Since then the two subspecies have largely remixed, blending their genetic heritages and thereby producing rather unique local populations called *Mus musculus molossinus*. This recombination of heredities, referred to as reticulate evolution, does indeed exist in nature.

The Mice of Milan. Intraspecies subdivisions may be even more complex. The domestic mice of Italy, for example, exhibit an interesting phenomenon concerning the number of their chromosomes. The normal number of chromosomes in the

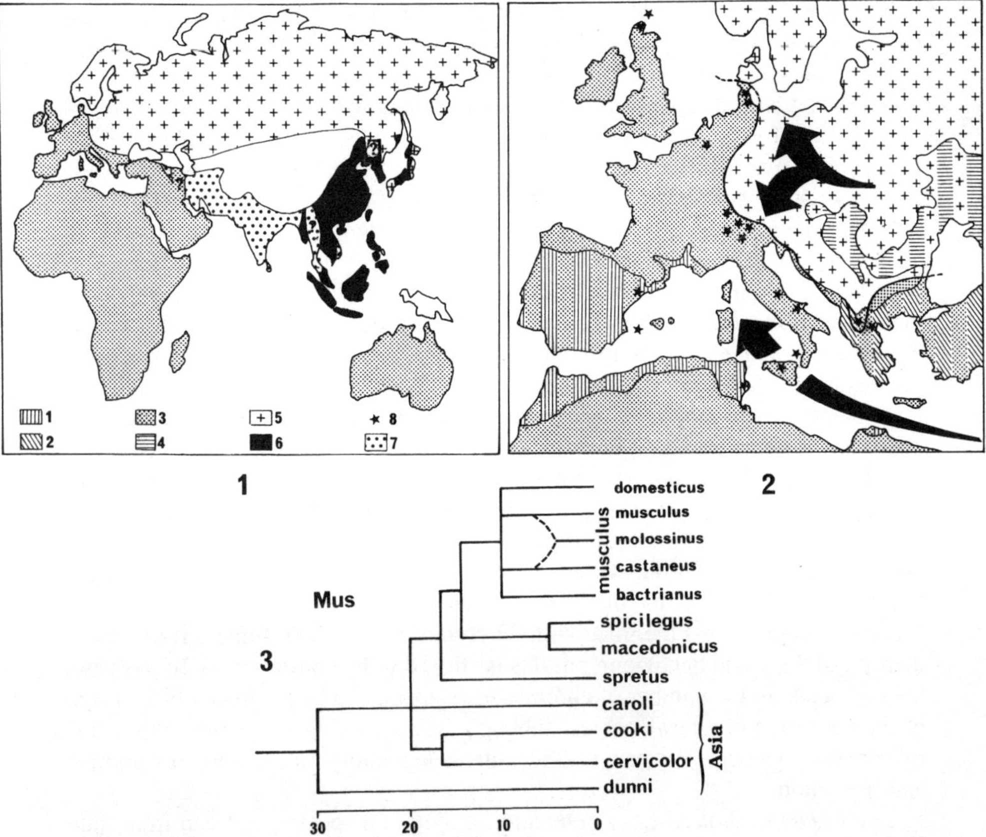

Fig.7.1. Subspecies of the domestic mouse. Studies of the genetic variability and biochemical characteristics of these animals facilitate the identification of mice on the continents. It appears possible that the home of the common ancestor is found somewhere in the southern part of Asia. In a succession of migrations, mice have colonized virtually all parts of the continents, especially by following man in recent times. The domestic mouse *(Mus musculus)* is the most widely distributed form. **1** Distribution of the main mouse species and the subspecies of the domestic mouse (*1 Mus spretus; 2 Mus macedonicus; 3 Mus musculus domesticus; 4 Mus spicilegus; 5 Mus musculus musculus; 6 Mus musculus castaneus; 7 Mus musculus bactrianus; 8* mouse with transformed karotype). 2 Distribution of subspecies of the domestic mouse (symbols as in *1*). The *arrows* indicate the direction of migration. *3* Genealogical tree of mice as constructed from genetic differences observed by electrophoresis of soluble proteins (differences grade from 0 to 30). the tree is based on the assumption that the genus *Mus* appeared with a standard karotype with 40 chromosomes. The subspecies *Mus musculus molossinus* represents a hybridization between two earlier established subspecies and is thus a rare example of reticulate evolution. (After Bonhomme and Thaler [3].)

mouse is 40. However, the mice living in the city of Milan possess only 24 chromosomes and in the outskirts of the town, populations of mice are encountered with chromosome counts ranging from 24 to the normal 40. All these mice are able to interbreed with one another and appear to belong to the same species. This observation is of great evolutionary significance. Some specialists maintain that the formation of a new species implies a change in the number of chromosomes. this is certainly not the case and, to the contrary, a great difference in the number of chromosomes does not directly prove the separation of two species. It was possible to show through electrophoresis that all Italian mice have the same genetic constitution. This would imply that the evolution of the chromosomes took place only after the arrival of mice in Europe, i.e. less than 10000 years ago . It must have occurred more rapidly than that of the proteins involved as these have remained unchanged. This example shows clearly that there is also a hierarchical ordering in evolutionary changes which may take place quite independently of each other, sometimes affecting the gene level and sometimes the chromosomal level.

The Mice of Kairouan. The domestic mouse which we have discussed so far also occurs in Tunisia where a race with 22 chromosomes is found in the city of Kairouan. In the suburbs and surrounding countryside, this race is replaced by the normal mice with 40 chromosomes. So far no hybrids between the two forms have been observed, suggesting that the two races almost behave like two distinct species. It appears that the mice with 22 chromosomes show some differences in their physiology and behaviour, but this is still awaiting confirmation. In this case, the difference in the number of chromosomes appears to be the most obvious sign of the formation of a new species, although from a biochemical point of view the two are still identical. Mice are thus a veritable laboratory for the study of variation and speciation.

The Cabbage Butterflies of Scandinavia. Within a species, one can frequently observe that the gradual variation of a certain feature parallels the "continuous" change in an environmental factor such as temperature, humidity, or altitude. Such variations are referred to as "clines". In Scandinavia, the pigmentation of the upper surface of the wings of cabbage butterflies becomes increasingly darker from the southeast to the northwest.

Males and Females. Another type of variation within different populations of a species is the presence of several morphologically different, discontinuous groups: the so-called polymorphism within species. The most obvious example of such a difference is that between males and females: so-called sexual dimorphism. In some cases males and females of the same species have been classified as entirely separate species.

However, in some other species, such as certain invertebrate molluscs, polymorphism will lead to the establishment of several well-defined forms. For the central European snail *Cepea nemoralis,* the individuals from warmer, drier habitats will show yellow unbanded shells whereas those from cold, humid climates exhibit red banded shells.

The origin of Species

The formation of a species or, to use the "scientific" term, speciation corresponds to a complex assembly of processes and mechanisms. The problem of the origin of species is one of the key problems to understanding biological evolution. This was clearly perceived by Darwin who entitled his fundamental work *On the origin of species* [7]. The various models of speciation that have been proposed by biologists have called upon the intervention either of external factors like geographical isolation or of internal phenomena like divergence. Specialists have shown a tendency to extrapolate from a certain mechanism for speciation to the rest of the living world. However, as was aptly pointed out by E. Mayr, there is a multitude of processes for speciation, and it depends on the particular circumstances, the groups involved, and the specific environment.

Speciation by Geographical Isolation

We are dealing with speciation by geographical isolation when there is a division of the area of distribution of the ancestral species into several parts. This leads to the establishment of at least two geographically isolated groups of populations.

The Dumbbell Model of Geographical Speciation
The classic example for this is a widely distributed species which becomes split into roughly equal parts by a geographical barrier. The formation of two new species is then envisaged as resulting from the progressive divergence between the two populations due to the interruption of the gene flow, which would otherwise have maintained the uniformity of the species. This model is also known as the symmetric bell model (Fig. 7.2).

The Asymmetric or Peripatric Model of Geographical Speciation
The establishment of geographical species mostly results from an asymmetrical fragmentation of the ancestral species' area of distribution into two areas of rather different size. The size of the peripheral, isolated population is a factor which influences the formation of a new species. It may range from an isolated population of several individuals to the extreme of only one remaining individual (pregnant female) which is then referred to as the "founder individual".

This asymmetric model was developed by E. Mayr [8] (Fig. 7.3) from observations on the Indo-Pacific archipelago. When considering a series of bird species which are related but geographical separated, one will always find that those farthest removed from the parental species are the ones most distinct from it. Sometimes they have even been classified as belonging to a different genus altogether. This observation led E. Mayr to propose the so-called peripatric model.

The reduced size of the founder population, which induced certain English-speaking authors to use the term "bottleneck", is one of the factors which would explain some of the particular phenomena that lead to the isolation of a new

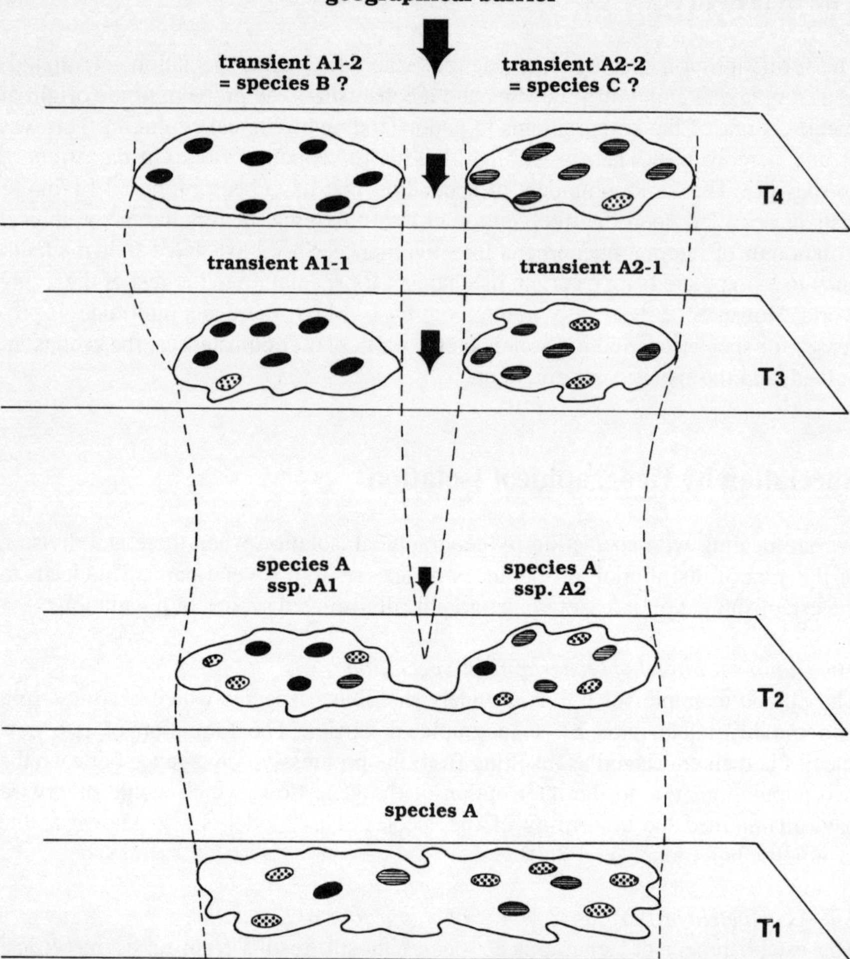

Fig.7.2. The symmetric model of speciation (dumbbell model). A species *A* at time *T1* is present over a wide area, exhibiting wide geographical variability; the circles indicate populations with different genotypes. At *T2,* a geographical barrier starts to develop, splitting the area of distribution into two subareas with an approximately equal number of individuals, but without completely interrupting the genetic flux. The two groups of populations then start to diverge, differentiating into the two subspecies *(ssp) A1* and *A2* of species A. At time *T3,* the geographical barrier (sea, mountain range, forest, glacier, etc.) completely isolates the groups of populations from each other; the populations then evolve independently and gradually, leading at *T3* to the transients *A1-1* and *A2-2* (populations in temporary transition). At *T4,* divergence has advanced to the degree where transients *A1-2* and *A2-2* no longer resemble the ancestral species. Have these transients now given birth to the new species *B* and *C* respectively? For this to occur, the two transients *A1-2* and *A2-2* would have to come into contact, so that we might find out whether their divergence is already advanced enough to stop them from interbreeding. Should interbreeding have become impossible, then divergence has led to reproductive isolation and, indeed, to the establishment of two new species. This mode of speciation is referred to as balanced or symmetrical because of the separation of the population into two equal parts. It has been observed in a number of fossil cases, but appears to be rather rare. (After Chaline [17].)

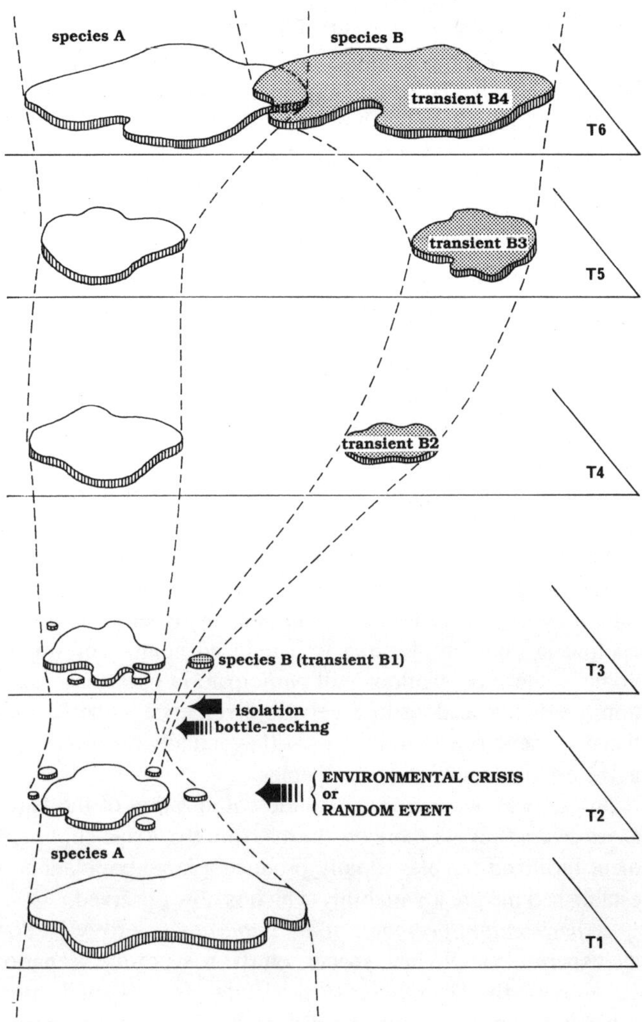

Fig.7.3. The "founder" population model of E. Mayr. Species *A* at time *T1* is in equilibrium with its environment. At time *T2,* an environmental crisis or any other random event has led to the formation of peripheral isolates. A reduction of the number of individuals, genetic drift, or inbreeding may unbalance the genetic programme of A, leading to the new programme B which will cause B to establish itself by reproductive isolation from A. This eventually results in the new species *B*. Species A will suffer fluctuations in the number of its individuals and geographical redistribution from *T3* to *T6* without, however, suffering any modification of its genetic programme. From time *T3* onwards, species evolves gradually and at any time between *T3* and *T6* it is present as a transient form, a temporary transitional population representing a state or degree of evolution within the lineage B. At *T6,* species *B* extends its area of distribution, coming into contact with its source species *A* with which it is now able to live alongside but not interbreed. This coexistence without interbreeding is the first real test of reproductive isolation acquired at time *T2*. It did, not however, have any significance before species B came into contact again with its source species A. This mode of speciation appears to have been followed most widely during geological times. Frequently, the parental species A disappeared when it came into contact with the daughter species B, as the descendant species is usually more competitive than the parental species. (After Chaline [17].)

species. It will entail a considerable loss of genetic variability as only a small portion of the genetic heritage of the species will be represented. Such small marginal populations will frequently find themselves under rather limiting conditions, close to different ecological niches which they would be able to occupy. These isolates are prone to what is known as a "genetic revolution", a term which only masks our ignorance of the processes involved. In these small populations there are generally random variations in gene frequencies, the famed "genetic drift". Certain rare genes may suddenly become dominant or completely eliminated. Furthermore, the reduced size reinforces the possibility for consanguineous breeding which would increase the chance of new features becoming established. While these isolates are usually destined to become extinct, the existence of new species in the palaeontological record proves the occasional success of speciation by a bottle-neck.

An Example of "founders" – the Hawaiian Drosophilas. Harryton L. Carson [9] has proposed a model for speciation from a single founder individual that is capable of producing numerous sexual descendants. This would have to be a fertilized female arriving in a new environment where there is no competition from similar species.

Carson's model of speciation invokes an alternation of successive phases of expansion and reduction in number. He proposed that the first phase of expansion, saturating the new habitat with organisms, is succeeded by a phase of abrupt reduction in numbers, leading to a reduced number of small isolated populations. Some of these populations will participate in new phases of expansion and reduction, gradually leading to a genetic divergence – the famous but still unsubstantiated "genetic revolution". This will eventually result in reproductive isolation and in the establishment of a new species.

In this way we can visualize the colonization of the Hawaiian island chain by *Drosophila,* brought there by storms from the Asian continent. In the new environment fertilized females rapidly produce a large population, and further mutations established the great variability which is now observed.

Experimental Speciation in the Laboratory. Jeffrey R. Powell [10] tried to create experimentally a new species on the basis of the scenario proposed by Carson. He allowed the *Drosophila* to proliferate from a small group of individuals and subjected them to several successive phases of expansion and reduction in number: four groups of *Drosophila pseudoobsucra,* each made up of 20 males and 20 females, were placed in cages under conditions permitting rapid expansion. After maximum expansion, he artificially reduced their number by selecting 12 couples to form 12 founder populations to initiate a new phase of expansion. After 15 generations, tests of reproductive isolation showed that of 45 combinations between individuals selected from the various populations, only one had developed some degree of behavioural isolation. Thereafter, eight of the 12 populations were subjected to three further cycles of expansion and reduction. Fifteen generations after the last reduction in quantity, three of the eight populations exhibited behavioural isolation in the pre-mating stage which persisted from then onwards. No post-mating isolation could be observed. None of the control populations, which were not subjected to these cycles of expansion and reduction, showed any signs of reproductive isolation.

An Irritating Problem: Chains of Subspecies. Divergence between subspecies is frequently interpreted as one way of leading to the formation of a new species. A classic example of the divergence among subspecies is presented by the warblers which occur in five continuous subspecies along the Himalayan mountain range (Fig. 7.4.). The successive subspecies are able to interbreed, but at the end of the chain where the two extreme subspecies come into contact and live alongside each other, there is no hybridization. These two subspecies behave as if they were two distinct species. This observation shows that divergence, regardless of the level at which it may occur, tends to accumulate step-by-step along the chain of subspecies, eventually leading to reproductive isolation and thus to the establishment

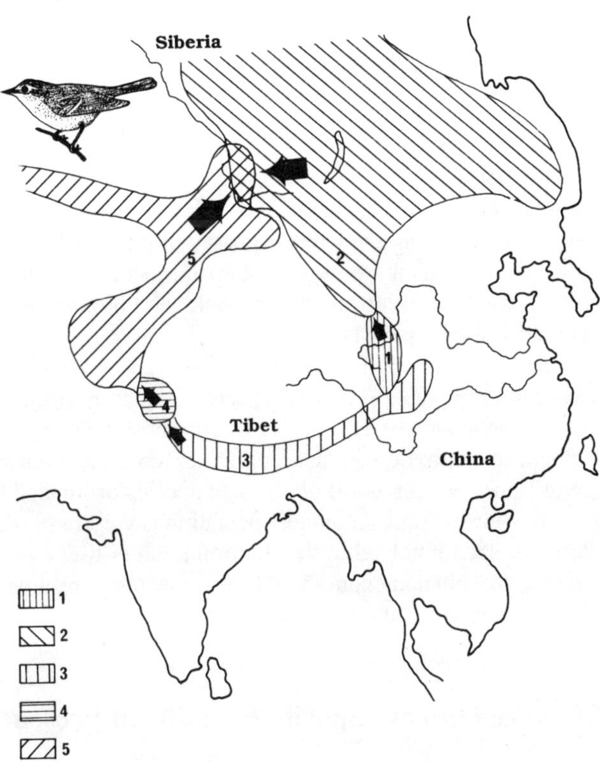

Fig.7.4. A chain of subspecies: the Himalayan warblers. The warbler *Phylloscopus trochiloides* is widely distributed throughout the mountain ranges around the arid plateaus of Central Asia and in northern Asia. Two subspecies coexist in the Altai mountains *(2 plumbeitarsus; 5 viridianus)* without being able to interbreed with each other. However, the two forms are related to each other through a chain of subspecies in which the adjoining ones are able to interbreed with each other *(1 obscuratus; 3 trochiloides; 4 lludlowi).* The area of encounter of the two non-interbreeding subspecies is undoubtedly of fairly recent origin (postglacial). The example illustrates that the cumulative effects of divergences between adjoining subspecies may lead to reproductive isolation. If this chain of subspecies were to be interrupted by some event, two distinct specific groups would form. From the point of view of nomenclature, this presents a nearly insoluble problem for a systematist. (After Ticehurst [18].)

of a new species. This provides a headache for systematists who would like to give a name to each of these forms, but is a most remarkable example for students of evolution as it proves the possibility of speciation from chains of subspecies.

Speciation Without Geographical Isolation

This theory was supported by geneticists, and in particular by H. de Vries [11] and Walter Bateson [12], who adhered to the typological concept of a species always reproducing itself without change. Mutations are then considered as a fundamental reorganization of the original type which could eventually lead to the creation of a new species. This approach was revived by R. Goldschmidt [13] and O. Schindewolf [14] who advocated the possibility of speciation from the formation of a "promising monster". However, monstrous mutations are generally lethal to the individual. We now know that mutations result from errors during DNA replication and that they do not have profound immediate implications for the organism, as they are frequently neutral.

In actual fact, there appears to be no need to resort to drastic mutations to explain the appearance of major new morphological types. In particular, the chronological shifts during the various phases of development provide a mechanism and processes by which we may explain the occurence of very large morphological changes that result from minor genic or chromosomal modifications (see the "horned wolves", p. 121).

The instantaneous formation of a species is a reality and is especially widespread among plants. The new species frequently results from polyploidy, the abrupt multiplication of the number of chromosomes. In such cases, the chromosome formula of the hybrid is made up of the combined chromosomal complement of the two species. One-third of all plant species originated by this process. Amongst animals, however, this mode of speciation is very rare, mainly because of the complex processes involved in the determination of the sex of the descendants. Consequently, the phenomenon should be more frequent among animals in which the eggs develop without fertilization.

May We Observe Species Forming at Present?

For the biologist A. Vandel [15], evolution had achieved its aim with man. For P. P. Grassé [16] the formation of the major body plans constituted the essence of evolution (Chap. 10), to the extent that recent evolution, reduced to the formation of new species within these plans, appeared to him as a simple secondary, but continuing, process of elaboration.

Evolution, however, is still operating as is illustrated by the divergence presently taking place within certain species. We have given several such examples with *Drosophila* and rodents and, in particular, with the domestic mouse (Fig. 7.1.).

A Fundamental Concept, the Hierarchy of Organizational Levels in Organisms, the "Keyboard of Living Matter"

Through an environmental crisis or a random event, the area of distribution of a species becomes modified, fragmented into small peripheral isolates, or even broken up into a mosaic of isolated minute populations made up of only a few individuals. The isolates situated around the edge of the area of distribution of a species find themselves frequently confronted with rather limiting environmental conditions, close to new areas which they have not explored. As a result, they may exhibit extreme morphological variation. The majority of such isolates will be condemned to disappear. The small size of the populations will permit genetic drift and consanguineous inbreeding to act on a reduced genetic programme, leading to a reorganization within the species. At which level will this reorganization take place?

The answer is now quite clear: it may occur at any level. At the genic level there are mutations, gene recombinations, and regulations, whereas at the chromosomal level chromosomal mutation will modify the arrangement of the genes, their mutual separation, and therefore the various types of interaction between them. At the biochemical level new proteins or enzymes may be produced, while at the physiological level modifications of the functional systems of the organism and its metabolism may occur. At the developmental level changes in the programmes which control development (chronological shifts) may lead to morphological differences of varying importance. At the ecological level, either the niche occupied by the species or the reproductive strategy may be changed. And, finally, it may also be changes at the behavioural level as we have seen in the examples from mice and *Drosophila*.

Some of these modifications will lead to important secondary effects, notably the prevention of interbreeding between the parental species and the isolates if contact is renewed. Modifications may contribute to eventual reproductive isolation, for example, alteration of the mating organs, changes in the timing of sexual maturity, changes in reproductive behaviour or aggressiveness, etc. Other changes could lead to the sterility of the descendants by chromosomal modifications which prevent the normal multiplication of cells or by incompatibility of the chromosomes during formation of the sex cells. Modifications may occur prior to or after mating.

The analysis of divergence between daughter species and their supposed ancestral form shows that the divergence may be quite different between one species and the next, or between one level and the next. In one case there may be a chromosomal change, and in another instance there may be an ecological or behavioural modification; it is also possible that there may even be changes at all levels at the same time. There appears to be some independence between the organizational levels of living matter. Modifications at the basic gene level do not allow us to predict or outline changes at higher levels, which frequently exhibit entirely new properties.

We are dealing here with a fundamental point and can suggest that the formation of a new species may be compared to the score played on an organ that is made up of as many different keyboards as there are organizational levels in living matter. These levels are arranged in a hierarchy from genetics to behaviour and incorporate en route an increasing number of possibilities. Eventually, the "score" for the formation of a new species may be played at independent levels, and does not necessarily require the entire hierarchy of keyboards. For simplicity we shall refer to this concept as the "keyboard of living matter" or, more scientifically, as the hierarchy of the organizational levels of the living world.

Modifications within an isolate can thus lead to reproductive isolation which will, however only become notable when the modified isolate regains contact with the parental species. If the populations remain geographically isolated and have no chance of encountering each other again, then reproductive isolation no longer serves a purpose and makes no further sense.

From a purely scientific point of view, reproductive isolation is a secondary effect that results from modifications which may be favoured by a specific situation of the reduced number of individuals in the isolates. From the evolutionary point of view, however, it is a rather fundamental concept. The role of reproductive isolation is that of a barrier preventing the new genetic programme from remixing with that of the source species when two populations which have descended from the same source come into contact with each other. Once reproductive isolation has been demonstrated, the birth of a new species is proven.

Contrary to popular belief, evolution is not yet finished and there are a number of species presently forming in the world arounds us. However the formation of these species frequently requires many generations and as we do not immediately know which ones will result in new isolates, new species will not always become evident within the relatively short observation period available to biologists. Time, an essential dimension of speciation is the privileged domain of palaeontologists but this will allow us to outline in the following chapter the formation of a species in its three dimensions: namely, internal, spatial, and temporal.

Chapter 8

The Historical Framework of Evolution

One of the first difficulties encountered when trying to understand evolution is the concept of geological time, for which we have no comparison in everyday life.

What Are Time and Space for Biologists and Palaeontologists?

Biologists study the forms presently living on earth, analysing their structures with the aim of eventually placing them into the well-ordered inventory known as systematic classification. The biological approach entails certain aspects of molecular constitution, genetics, sexuality, physiology, development, biology, ecology, and behaviour, to mention only the most important aspects. Biologists attempt to elucidate processes, i.e. a whole array of active phenomena organized in time, that allow the advance from the genetic programme through development to the adult individual. Biology is also interested in mechanisms (physiology), i.e. the working of the organs which facilitates the overall harmonious functioning of the organism. This may be perceived as being rather like a well-tuned machine. It is frequently only through studies of the irregularities in the operation of these machines that the processes themselves may be understand. All these approaches are combined in the study of speciation; however, in this study the biologist will have to concede the limitations of a specific approach which only covers an instant in geological time. In the formation of species, as we shall see later, the one important factor is succession of a great number of generations. This is why biologists studying these problems usually select as their material extant species which reproduce rapidly. They prefer organisms with a rapid succession of generations such as *Drosophila,* in which the cycle time is only about 15 days. Biologists lack the essential factor of biological evolution, namely, time.

The dimension of time is the privileged realm of the palaeontologist who studies the inventory of extinct species, attempting to establish ancestor-descendant relationships. During these analysis the palaeontologist studies the circumstances and peculiarities of the historical advance of evolution – the very patterns of evolution. Thus, palaeontology is the science of the history of life which should be placed within the framework of the history of the Earth, and we know from the theory of plate tectonics that continents and oceans have a complex history which, to some extent, has also influenced the history of life on the planet.

The concept of time employed by palaeontologists is rather different from that used by biologists. The palaeontologist is accustomed to thinking in thousands, millions, and even billions of years, whereas the biologist can, at best look back over only a few generations. It is difficult for us to comprehend such long periods of time as our individual life spans are counted only in years, and that of a country or civilization is only in millenia at best.

A Thought Experiment

The concept of time to a geologist is like that of distance to an astronomer. As distances are easier to comprehend than the much more abstract concept of time, we shall try to illustrate periods of time by comparing them to astronomical distances. To do this, we shall undertake one of the thought experiments recommended by Einstein, an experiment which does not require a large budget. Let us imagine that an observer is on a planet circling a distant star and asks us the following question: Taking into account the speed of light at 300 000 km/s or about 9460 billion km/year 9.46×10^{12} km/year, to what distance would I have to go to be able to observe the appearance of man on earth some 2 million years ago? Or to see the start of the Tertiary, about 60 million years ago? Or the start of the Mesozoic 225 million years ago? Or the beginning of the Palaeozoic some 600 million years ago, or even farther back to the appearance of life itself?

The star closest to our solar system, Proxima Centauri, is 4.2 light years, or 39 735 billion km, away and appears to us as it did just over 4 years ago. An inhabitant of that star would view the earth as just over 4 years and 2 months old.

To go farther back in time, the Andromeda nebula is some 2.25 million light years, or 21 billion billion km (21×10^{18} km), away. An observer there would now be witnessing the first appearance of man on earth.

The Virgo cluster of galaxies is about 50 million light years, or 473 billion billion km (4.7×10^{20} km), away and from there one would see the Earth as it appeared at the beginning of the Eocene epoch, when the mammals were starting to spread.

From the Perseus or Pegasus clusters, some 1892 billion billion (1.9×10^{21}) km away, one could see the break-up of the continents and the appearance of the mammals during the Triassic, some 200 million years ago. From the Great Bear, 3122 billion billion (3.1×10^{21}) km away, one would find the Earth in the Carbonifeous epoch, some 300 million years ago. From Centaurus or the radio galaxy Cygnus A an observer could see the Earth as it appeared at the beginning of the Palaeozoic when the invertebrates were starting to spread, some 600 million years ago.

The last two astronomical objects are about 5676 billion billion (5.7×10^{21}) km away.

As Hubert Reeves aptly put it: "To look further back in time, we have to look farther out" [1]. To witness the appearance of the first blue-green algae (cyanobacteria), or to observe the first anaerobic bacteria in the primordial soup, we would have to intercept light that originated from the Earth more than 3.5 billion light years ago, now some 3.1 x 10^{22} km away.

And if we were to observe the formation of our solar system and the Earth with it, we would have to go back about 4.5 billion light years to the most distant galaxies known so far, such as the galaxy 3C-295 Bouvier. This is about 47 304 billion billion (4.7 x 10^{22}) km away.

This exercise serves to illustrate that the history of life is a phenomenon unfolding on a cosmic scale, the duration of which is as awe-inspiring as the Universe itself. Let us now go back to the first stages in the history of life.

The History of Life

In order to be complete, a theory of evolution will also have to address the problem of the very origin of living matter. Up to the time of Pasteur, the concept of spontaneous generation was accepted. Just as flies were believed to form out of putrefying meat, or bacteria to appear in culture media, so the possibility of the immediate complete formation of complex organisms was accepted. However, Pasteur was able to show that under controlled conditions any bacterium resulted from an earlier bacterium; he thereby shattered the concept of spontaneous generation. But, with this materialistic approach, the problem was only bushed back to the distant geological past of the planet because, initially, life would have had to form spontaneously from inorganic substances.

It was Alexander I. Oparin of the USSR and John Haldane of Great Britain who formulated this question of a materialistic evolutionary basis. The origin of life thus became the object of scientific study. What a tremendous intellectual challenge!

We shall have to divide this history into several stages: To start with, the formation of living material, the chemical phase, was followed by the biological phase of assembling these materials into a living system or, in other words, a cell, of which the present bacteria give us some idea, but which became further transformed by more than 3 billion years of evolution. Eventually, this cell was transformed to become what we now call a eukaryotic cell (Fig. 1.1); such cells make up the majority of all unicellular and multicellar organisms.

At first sight this evolution, which extended over some 2.5 billion years, appears to be interminable. However, simple as it may appear to us, let us consider the complex system of genetic material in a bacterium, and the processes which translate the information into proteins. These processes involve biochemical activities which build up the cell substances that allow the bacterium to recognize, albeit in a rudimentary fashion, the conditions of its immediate environment, and allow it to respond to them adequately. And finally, as pointed out by F. Jacob: “what seeks to produce without respite, a bacterium, are actually two bacteria. Well, this appears to be their only aim, their only ambition.”

The plan of the eukaryotic cell was much more ambitious, as it developed sexual reproduction and built multicellular organisms; these were major evolutionary events resulting in an explosion of diversification in the organic world.

The first known multicellular organisms made their appearance only about 680 million years ago. Then, over a very short time span, multicellular organisms achieved their entire diversification, so that evolution advanced very rapidly.

Our understanding of each of the various phases shows considerable differences. For the first stage, we can formulate a number of hypotheses which can be discussed and, although they may need to be revised, are at least based on experimental data. Concerning the appearance of the cell, we can only understand a very few aspects. Precambrian terranes, from 3.5 billion to 600 million years provide old, only scant remains of the unicellular organisms outlining the first biological evolution. However, our documentation of this period has improved greatly over the last few decades. The final stage, that of multicellular organisms, is the one best demonstrated by fossils. Gaps in the record are still numerous and while some of them could be closed, others will most probably have to remain open.

The Origin of Life on Earth

Phase One: The Constituents of Living Matter Are Formed

The planet Earth, formed some 4.5 billion years ago, offered favourable conditions for the formation of living matter. It is neither too close to the sun, and thus is not too hot, nor too far away from it, and thus is not too cold. Its surface temperature permits water to remain a liquid. The Earth's mass is sufficient to retain an atmosphere, although the original composition was notably different from that at present (Appendix 8.1).

The atmosphere, liquid water, energy supplied by the sun, lightning, and volcanoes were the reagents of this first chemical phase. The simple molecules of the atmosphere became assembled into more complex molecules. When dissolved in water, they formed still more elaborate structures which are called molecules of "biological interest" as they are the essential ingredients of living matter. These molecules would have accumulated in shallow water or close to the mid-oceanic ridges [2, 3, 4, 5, 6].

This tentative reconstruction of the first phase is not a mere hypothesis, but is based on the results of experiments carried out in an apparatus devised by Miller (Fig. 8.1). From this we can draw the remarkable conclusion that organic molecules may form from non-living matter or inorganic molecules.

Phase Two: The Materials Are Assembled in to Cells

We shall now consider the problem which is most difficult to comprehend: By what process can the organization of a cell emerge from unorganized matter?

We may define life by, in addition to other characteristics, its possession of a message, the genetic code (DNA), and mechanisms by which this message is decoded for the synthesis of proteins. Viruses, the simplest biochemical structures

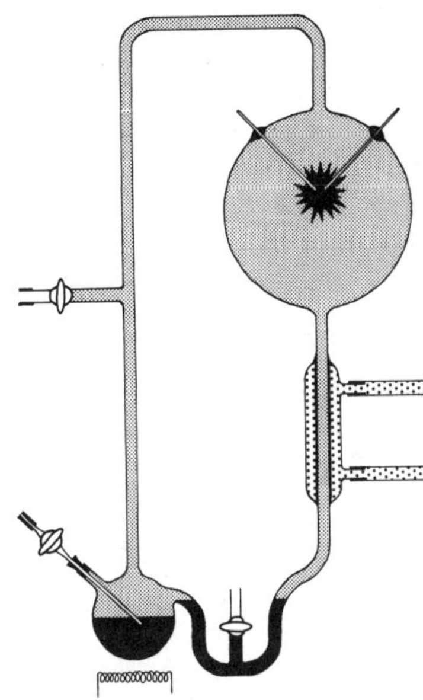

Fig.8.1. The Miller apparatus. The apparatus permits the simulation of atmospheric conditions of the early Earth. A mixture of the gases thought to have been present in the primitive atmosphere is introduced via the tube (*to the left*) of the rising duct. The water in the spherical container (*bottom left*) is kept on to boil, to keep the gases circulating in the apparatus. In the glass sphere (*to the right*), the gases are subjected to electric discharges simulating the sources of energy in the early atmosphere (thunderstorms, etc.). The gases are then cooled down in a condensation unit, with the molecules formed accumulating in the liquid at the bottom of the apparatus. By this method most of the molecules of biological interest , like amino acids, sugars, certain fatty acids, etc., have been synthesized. (After Miller and Urey in [4].)

known (DNA or RNA plus proteins) possess the genetic material but do not have the mechanism with which to decode it. To do this, they have to enter a cell as parasites. Therefore, a virus can not provide us with a model for the first living assembly.

The first cellular organization, the palaeocell, would have been much more complex. How could it assemble itself? Certainly not purely by chance. We may just as well try to reconstruct a watch by pouring its parts into a bucket and stirring, the complexity of such a watch is far removed from the complexity of a cell.

We shall explain some of the problems inherent in the formation of a cell. In a modern cell decoding of the DNA passes through the stage of transcription (Fig. 3.2.) which requires the participation of an enzyme, in this case RNA polymerase. Like any protein, it has to be coded by a gene which, in order to become decoded, would have to be transcribed by the very polymerase itself. Thus, when a protein or enzyme is to be formed, it has to be present already. We are now left with the classic problem of the chicken and the egg. Are there other methods of synthesis that do not require enzymes which may have existed in the palaeocell? How was this genetic code established and what were its mechanisms for transcription?

The 3.8-billion-year-old Isua rocks of Greenland may have contained organogenic carbon; however the first known organism is the bacterium *Eobacterium* from the 3.4-billion-year-old Archaean of South Africa. In Australia, indications of organic matter have been found in rocks as old as 3.5 billion years. In the following text, to any references to dates or durations should be added the reservation "according to our present state of knowledge".

THE AGES OF THE EARTH

Age in billion years	Era	Main events
	CENOZOIC	Primates
	MESOZOIC	Mammals Reptiles
	PALAEOZOIC	Fish
0.57	PRECAMBRIAN	
1		Eukaryotes
		Sexual reproduction
2		
3		Anaerobic bacteria
4		Oldest sedimentary rocks
		Formation of oceans and continents
4.5		Formation of the Earth

Era	Stages		Age m.y.
	Quaternary	Pleistocene	2
CENOZOIC	Neogene	Pliocene	12
		Miocene	26
	Palaeogene	Oligocene	37
		Eocene	57
		Palaeocene	65
MESOZOIC	Cretaceous		141
	Jurassic		200
	Triassic		235
PALAEOZOIC	Permian		280
	Carboniferous		345
	Devonian		395
	Silurian		435
	Ordovician		500
	Cambrian		570
PRECAMBRIAN			

Fig.8.2. Simplified geological time scale. (After Odin [11].)

In the advance of the evolution of the living world, as outlined by the scanty fossil record of the Precambrian [7, 8, 9], two events played a decisive role (Figs. 8.2 and 8.3).

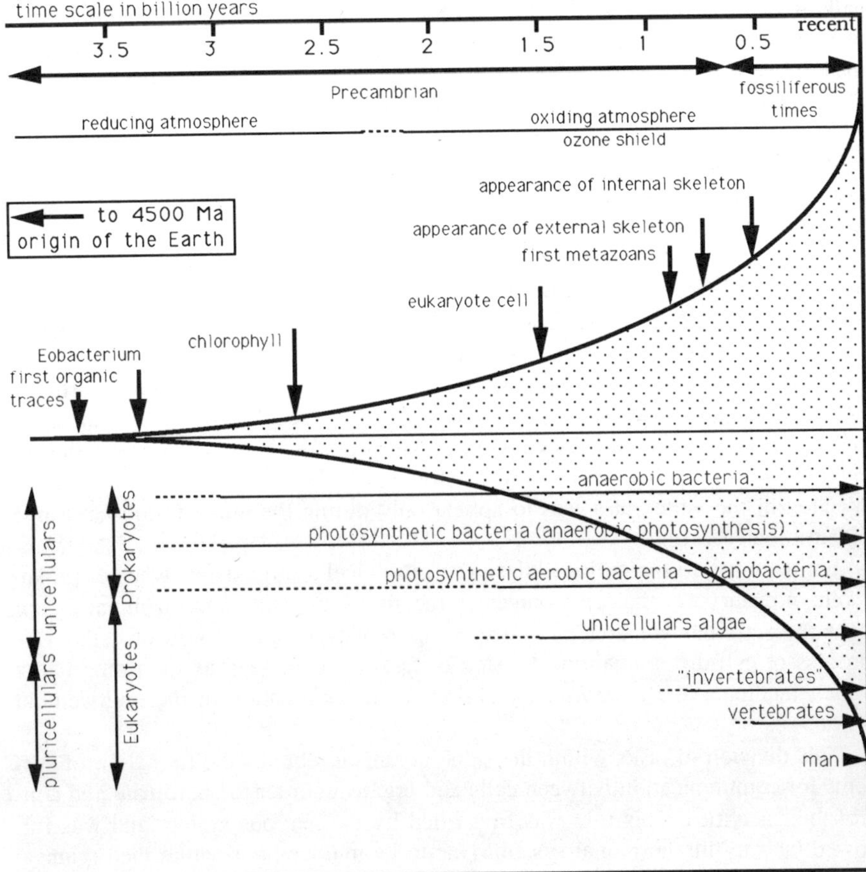

Fig.8.3. Chart of the history of life. The width of the *dotted area* does not imply any quantitative value, it only indicates that during virtually the entire Precambrian the degree of diversification in the organic world increased only slightly, until the truly explosive spread at the Precambrian-Palaeozoic boundary when the fossiliferous period started. The first metazoans of the Ediacara fauna of Australia (680 million years ago) comprise only soft-bodied animals like medusae, proto-arthropods (?), and prot-echinoderms (?), and especially organisms of rather enigmatic systematic positions. Formation of a skeleton developed at the start of the Cambrian, concurrently with the spectacular diversification of the various body plans. Most of the phyla known at present were already around then. Some phyla like the graptolites have since disappeared. The vertebrates apparently originated during the upper Cambrian, as suggested from probable bone fragments. True vertebrates are known from the Ordovician onwards. The first terrestrial (tetrapods) and land plants (mosses and fungi) date from about 400 million years ago. The first "reptiles" and phanerogames (plants with visible seeds) appeared 300 million years ago. Mammals and insects with complete metamorphosis are about 190 million years old, while birds are 140 and higher insects and flowering plants are 80 million years old. Appearance of reflective thought and articulated language developed about 2 million years ago.

In the 3.4-billion-year-old cherts of the Swaziland Supergroup, pristane and phytane, decomposition products of chlorophyll, have been observed, suggesting its appearance at this time. The appearance of chlorophyll led to the production of oxygen which was to radically change the composition of the atmosphere, eventually resulting in the present state. This event marked a turning point in the history of life: from here onwards, any further generation of living matter from non-living material was impossible; any of the chemical precursors of life reappearing in the atmosphere would have been immediately destroyed by oxidation.

The second event, which occured about 1.5 billion years ago, was the formation of the eukaryotic cell (Fig. 1.1.) which facilitated structural and functional novelties, like sexual reproduction, together with new evolutionary facilities, such as the formation of multicellular organisms.

The first of these multicellular life forms are known from the 680-million-year-old Ediacara Formation of Australia [10] and from other areas like Great Britain, Siberia, North America, etc. The life forms appeared as soft-bodied animals like medusae, "worms", and various enigmatic structures.

At the start of the Cambrian, 580 million years ago, the animal world virtually exploded into all its various phyla which, with some exceptions, persist to the present day. This explosion coincided with, or was provoked by, another event: the appearance of an external skeleton in the form of a shell or carapace. The internal skeleton of the vertebrates was to appear only during the upper Cambrian, some 500 million years ago.

Let us return to the multicellular event: in unicellar organisms, whether prokaryotic or eukaryotic, the cell assumes all the vital activities. In multicellular organisms the activities are divided between different types and groups of cells. This process of cell differentiation (division of tasks) can be seen as the prime factor which facilitates the extraordinary ability for diversification in the multicellular world.

This division of tasks within the same organism requires the formation of systems for communication between cells and organs in order to coordinate and control their actvities. This role was first filled by the nervous system and was followed later by the hormonal system. Due to its multicellular nature the organism will, by necessity, be complex. It cannot be just a simple aggregate of unicellular organisms. Nevertheless, we are led by logic to postulate that multicellular organisms resulted from unicellular ones. But how? Mystery still prevails.

During the course of their evolution, animals managed to gradually overcome certain environmental constraints. A first example of this was the ability of a number of invertebrates and all vertebrates to maintain in their interior milieu or blood a constant concentration of the dissolved salts of sodium, calcium, potassium, etc. Only the vertebrates succeeded in achieving osmotic regulation, which maintains the osmotic pressure of their blood during any fluctuations (see Glossary). And eventually, thermal regulation released birds and mammals from the uncertainities of environmental temperature.

Two more remarks are called for at this state. Having been the sole masters of the Earth for over 2 billion years, the unicellular organisms had at their disposal all the planet's resources and possible ecological niches, although their state of organization allowed them to occupy only a rather limited number of the latter. So that, multicellular organisms, once formed, found many niches to occupy and this is another of the factors responsible for their diversification into the numerous types of body plan, or phyla. Such a possibility did not present itself again, save for the proliferation of the vertebrate phylum which was the last major body plan to appear.

Throughout the course of history, extinction has occurred time and time again, as new, more complex animals have managed to succeed the less complex, earlier organisms. However, in most cases, coexistence is eventually established, concurrently with a redistribution of the niches. The sponges, which, since the Cambrian, have filtered water to obtain their nourishment, coexist in modern oceans with lamellibranch molluscs and whales which are also filter feeders, allbeit more recent and more complex.

Chapter 9

The Species in the Course of Geological Time

The history of life may be compared to a huge cable enclosing as many strands as there are fossil and present species. The biosphere, with its 2 million species, would then correspond to a transverse section through this cable.

What Makes palaeontological Species Special?

Palaeontological species exhibit special characteristics which are known only from fossils, and not from living organisms. Fossils usually represent the durable portions of organisms such as the skeleton, shell, or carapace, and only rarely are the soft parts preserved. Compared with living beings, they are therefore incomplete witnesses, and the palaeontologist has to operate like a detective, with the fossils representing "circumstantial evidence" upon which reconstructions are based.

Palaeontological and biological species have been widely discussed and compared. It is obvious that in death an organism loses a large amount of information contained in its genes, chromosomes, or embryo. What is lost irretrievably is its physiology and metabolism, soft tissues, and way of life, together with its relationship with the environment, and its behaviour. In most cases only the skeleton remains. This may appear rather little, but even with such restricted data palaeontologists are able to perform veritable miracles. The reason for this is rather simple: the skeleton of a vertebrate, for instance, is one of the fundamental elements of the morphology of this organism. Combining the skeletal characteristics with data on the urrounding sediments and on other species living alongside the one under investigation will lead to conclusions about the palaeo-ecology, that is the relationship between the skeleton and its environment. Palaeontologists also have at their disposal data on the various phases of development of the skeleton, information obtained from studying the growth of various individuals. They are thus able to reconstruct the paths of development, and their variations, for individual species. Furthermore, for most groups there are living representatives which facilitate an understanding of the majority of structures observed, and thereby about reasonably exact reconstructions to be made of the missing soft tissues. We are even able to recognize the existence of both sexes in ammonites and also in some echinoids, without knowing exactly which is the male and which is the female.

Palaeontological species are frequently referred to as ancient or palaeospecies. They are essentially morphological species (morphospecies), as they can only be analysed from their morphology. Specialists can evaluate the relationships between palaeontological species and living biological species. In the case of mammals, for which we posses a considerable amount of information, the palaeontological species are mostly known by their teeth, which prove to be most resistant, even under unfavourable conditions of fossilization. One could almost say that palaeontologists specializing in mammals tend to establish "tooth classifications" which describe "dental species". However, dental features have also proven to be a most reliable characteristic in current species [1].

It is clear, however, that palaeontologists will always be hampered in their studies by the existence of sibling species, which they will be unable to identify in fossil forms. Palaeontological species could, therefore, quite well be considered as collections of sibling species that differ in their ecological and ethological characteristics [2].

One of the problems for palaeontologists lies in the quantity of individual fossils. Remnants of some fossil vertebrates are extremely rare and some species are known from only a few skeletal fragments. Consequently, we do not posses a complete picture of the morphology of the individual, let alone the variability within the species to which the individual belongs. Where the palaeontological record yields a large number of remains we have to ask ourselves whether the fossils really represent populations belonging to a certain biological species.

With statistical methods palaeontologists have at their disposal a tool to resolve this question. As the individuals of a species are related to each other by their ability to interbreed, the variation of measurable features may be expressed by certain characteristic curves.

How are ancestor-descendant relationships between species established? One could assume that it would be easy to establish an evolutionary lineage or ancestor-descendant relationship between species. however, this is a very difficult task.

Within the framework of the synthetic theory, palaeontology has not yet adopted the necessary strict approach. One might expect that the succession of the palaeontological record would unveil, step-by-step and from one fossil to the next, the relationship between species, thereby permitting the establishment of evolutionary lineages. It should be admitted in, as it were, a dogmatic way that species evolve in a gradual fashion. When we observe in two fossiliferous beds that lie on top of one other but are separated by unfossiliferous beds, two fossils which differ in a number of features and apparently belong to two different species, we could postulate that there are intrmediate forms in the intervening beds in other areas. Only the poor quality of the palaeontological record might be held responsible for the local absence of intermediate forms. When establishing a lineage, the two species would be connected by a straight line or by a broken line used by more prudent authors, to suggest the hypothetical nature of this reconstruction. However, the situation is usually treated as though its gradual evolutionary nature had been proven [3].

This approach was also used for purposes of classification which, according to G. Simpson [4], are more a matter of personal opinion than of methodological soundness.

The entomologist Willi Henning [5] has to be credited with the introduction of a truely rigorous approach to establishing the closest relationships between lineages and phylogeny. This new approach has become popularized as phylogenetic systematics or cladistics. Based on the hypothesis of evolution, the cladistic method also attempts to establish ancestor-descendant relationships between species and the time of divergence of the lineages. It is based on the analysis and distribution of similarities between the species. These morphological similarities could be of varying significance, representing either a primitive feature that has been inherited from a common ancestor, the appearance of a newly evolved feature (innovation), or a similarity resulting from convergence.

The establishment of ancestor-descendant relationships between species and the other types of systematic subdivisions (genera, families, orders, classes, phyla) should be based on a distinction between common, newly evolved features and those derived from archaic ancestral forms. In other words, a rigorous approach to determining ancestor-descendant relationships (phylogeny) will attempt to describe truely homogeneous groups which are characterized by the same evolved features, and will attempt to uncover and delineate unnatural groupings which have been united by primitive ancestral features or as the result of convergence.

A cladistic approach should begin with current forms in order to determine natural groupings of increasing extent and of increasingly older origin. Ancestor-descendant relationships are illustrated by so-called cladograms in which the distribution of derived features is presented (Fig. 9.1).

One of the results of this critical approach was to show that there is no unique feature common to the reptile group and that the class of reptiles should therefore be abandoned. It is an unnatural grouping that should not retained in systematics. The same applies to the class of fish.

The cladistic approach thus represented a veritable revolution in phylogeny, although it is not entirely free from uncertainities, deviations, and excesses. It is not always easy to decide whether a certain feature is primitive or derived. In a number of cases recourse has to be taken to the analysis of development; however, the

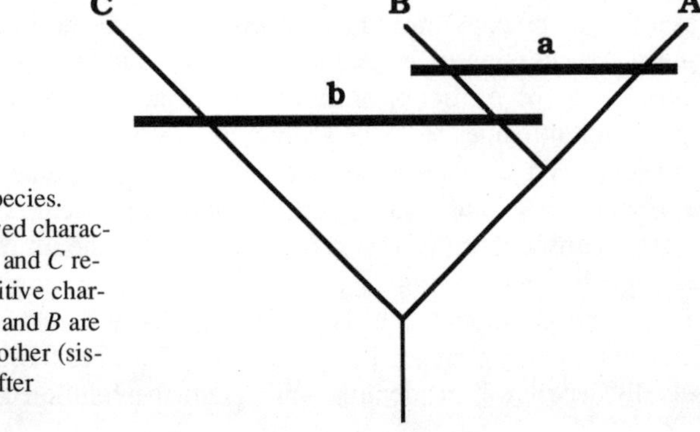

Fig.9.1. Cladogram of three species. The taxa *A* and *B* share a derived character *a* (apomorph) while taxa *B* and *C* resemble each other by the primitive character *b* (plesiomorph). Taxa *A* and *B* are more closely related to one another (sistergroups) than to taxon *C*. (After Goujet [36].)

complexity of the changes that occur during development demands that caution be exercised. Although cladistics represents progress in the analysis of characters, we have presented examples which show the limitations of the method. These limitations are all too frequently disregarded by theoreticians who particularly refuse to take into account the hard facts of the palaeontological record.

Species and the Continuity of Generations Through Time: The Lineage

When a new species becomes established by reproductive isolation, it will possess a new genetic programme. However, as we have shown for the domestic mouse, the species will continue to be modified over the generations from the genetic and morphological points of view. Throughout its existence, the species will be modified from generation to generation until its extinction. The progress or extension of a species over geological time will lead to the so-called lineage which may be subdivided into numerous subspecies living alongside each other across the globe.

The biological species concept described above only applies to modern species. To integrate the temporal or historical dimension of the lineages over geological time, we have to supplement the definition by E. Mayr [6] with a "temporal" component. This leads to the following spatial and temporal concept of a species [7, 8]: "A species is a homogeneous unit made up of natural populations succeeding each other in time, being at any moment of geological time able to interbreed with each other, but isolated from the other species."

We therefore have to consider a species as a dynamic entity, a historical phenomenon resulting from the succession of generations from its origin to its extinction. In this concept the biological species represents an "instant" in the evolution of a lineage. The problem we now encounter is the question of how to define a species in time.

A species is continuous with geological time, a situation referred to as the continuity of a lineage. This implies that from origin to extinction a species possesses a unique genetic programme which is modified with time. Because of this, we should refer to it by a name applicable to the lineage as a whole. The fossils discovered by palaeontologists at different successive geological levels will only represent instantaneous transverse sections through the lineage concerned. Considering the succession of generations and the genetic programme, any of these sections has to be artificial, or conventional at best. Nevertheless, palaeontologists give different names to species which represent sections of different importance in a lineage. What, therefore, is the value of these palaeontological species? They represent only evolutionary states within the same temporal species.

As a consequence of this concept, a species should retain the same name from its origin to its extinction. The palaeontological (or morphological species subdividing a lineage should only be considered as evolutionary stages of this lineage. The evolutionary changes in a temporal species are frequently quite important, especially when we are dealing with gradual evolution as in the case of the ammonites. Consequently, within the same lineage we would pass from one genus to another, and sometimes even from one subfamiliy to another. This method represents an outdated systematic approach which should be abandoned.

What Is the Life Span of a Lineage?

The life spans of species from birth to extinction are rather variable.

A new species may exist for only a short time when it comes into contact with another species occupying the same ecological niche more aggressively and with characteristics which render it more compatible with the environment in question (cf. domestic and field mice; Chap. 7). From the palaeontological record we know of well-documented cases of quite distinct species from adjoining deposits which disappear in younger strata; these species can have existed over only a few generations.

Usually the life span of a species will extend over a few million years. Gradually evolving species appear to have lived for 1–2 million years. However, species with pronounced morphological stability (stasis) have longer life spans, in the order of up to 10 million years. In this group we frequently encounter the so-called living fossils. One such organism is the small Triassic crustacean *Triops cancriformis* which appears to have existed for over about 200 million years without major morphological change [9] (Fig. 9.2). However, such stability is rather exceptional.

Frequently living fossils are only "closely related" to their ancestral forms and are not absolutely identical. the coelacanths are called living fossils as they curren-

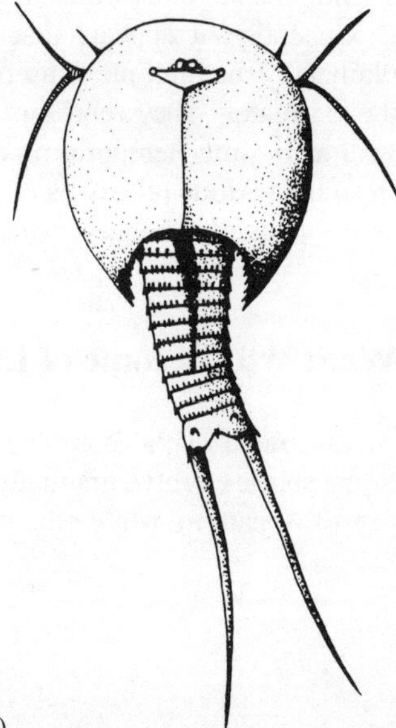

Fig.9.2. *Triops cancriformis*. (After Gall [9].)

tly exhibit fin structures which were in existence at a time when certain fish acquired limbs through a complex sequence of transformations. Therefore we may not say that the coelacanths have not changed for 200 or even 300 million years ago when these structures originated. They are exclusively modern forms, and although similar to the Jurassic examples, are by no means identical.

We also have to present the case of the nautiluses which for a long time have been considered as living fossils because of their primitive coiled external shell. Although the basic design has not changed, its potential has been exploited by the formation of numerous species [10] with repetitive morphologies, but without pronounced stasis. Consequently, we should talk of a type of particular longevity (panchronic) rather than of a living fossil, as the current *Nautilus* does not exactly resemble the Liassic, upper Cretaceous, or even Eocene forms. Furthermore, its true origin has not yet been fully determined.

A Special Model of Speciation by the "Bottleneck Effect"

Palaeontological records have shown a particular mode of speciation linked with the temporal dimension, the formation of a species by a reduction in the number of individuals with time (Fig. 9.3). When a certain species encounters difficult conditions, its area of distribution is broken up into a mosaic of residual isolates. Under these circumstances the phenomena taking place during speciation, as described in Chapter 7, may be initiated. These phenomena will not only operate in space, i.e. in the isolates situated along the edge of the area of distribution of the source species, but also in time. This is expressed by the term "temporal". Among the Jurassic brachipods of the Paris basin [11] this process must have operated more frequently than at other times in the geological record [12].

Generally, it appears that the origin of the majority of species could be explained by the multiple consequences of drastic reductions in the population size of these isolates. They represent bottlenecks in space (asymmetric spatial model) as well as in time (temporal model). When trying to compile an evolutionary model from the various processes of speciation and their spatial and temporal aspects, we arrive at a global concept which is outlined in the following paragraphs.

What Will Become of Lineages After Their Formation?

As illustrated by palaeontology there are at least three types of evolutionary mode. Some species evolve gradually in an irreversible manner, some exhibit a reversible repetitive pattern, while others appear to remain stable in their original form.

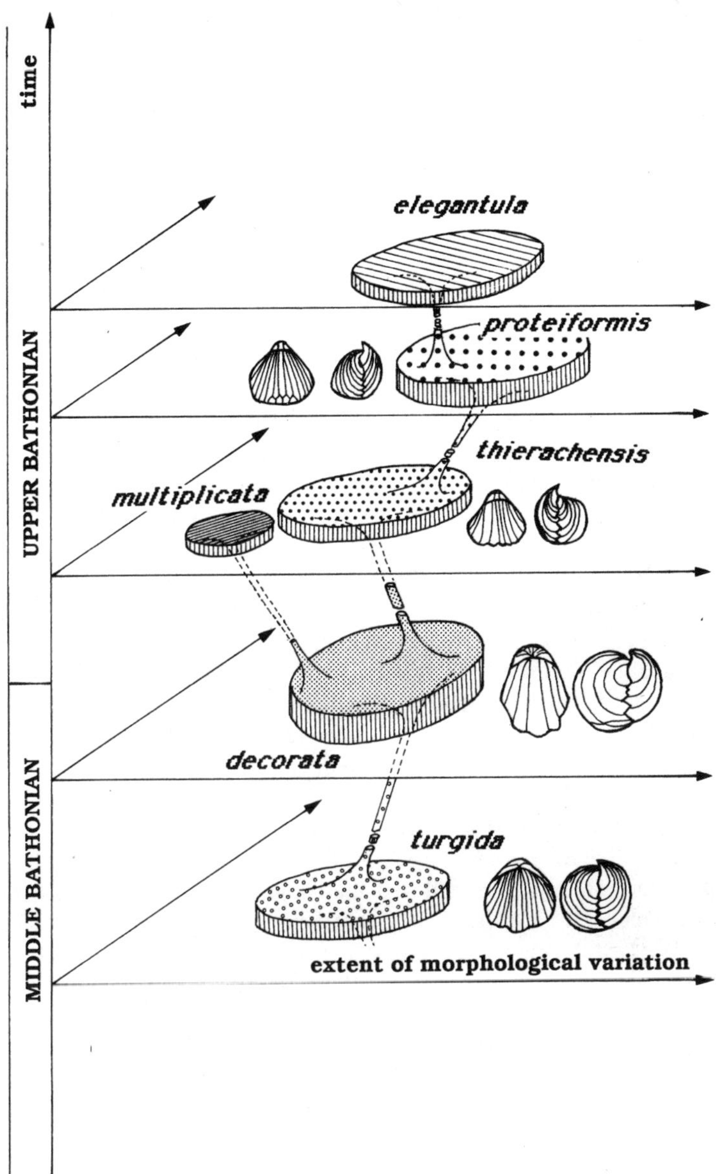

Fig.9.3. Model of speciation by the "bottleneck effect". The evolution of various species of the Jurassic brachiopod genus *Burmirhynchia* of the Paris basin. Under adverse conditions the number of individuals of the species was reduced drastically in the isolates. From these isolates, which retained only part of the genetic programme and were subject to genetic drift and the effects of inbreeding, new populations started to spread when the environmental conditions improved again. This type of speciation through a temporary bottleneck obviously occured frequently in the palaeontological record. In the present case, isolates of the species *turgida* gave rise to the species *decorata* from which, in turn, the species *multiplicata, thierarchensis, proteiformis,* and *elongata* formed. (After Laurin [11].)

Gradual Irreversible Evolution
The concept of gradual evolution was supported by Lamarck, Darwin, and the synthetic theory. However, geneticists like de Vries and Bateson advanced the opposite pattern; namely, discontinuous evolution. What really happens? Does gradual irreversible evolution really exist?

To prove the presence of gradual evolution very rigorous methods have to be applied. Any example should satisfy two criteria. First, a species should always be considered as a system of relationships between individuals. This requires the analysis of morphological variation within populations, based on a large number of samples. The second, more important point is that detailed palaeontological records must cover both the distribution in time (stratigraphy) and in space (geography).

Well-documented and thoroughly investigated examples are actually rare. Among the more recent studies, special reference should be made to microorganisms such as foraminifers [13] and radiolaria [14, 15], to ammonites [16–22], and to mammals [23, 24]. Amongst the latter, rodents in particular have been well studied [8, 25–28]. We shall describe two of these examples in this chapter and some other examples will be discussed in Appendix 9.1.

The Asian Foraminifer. One of the most instructive examples is shown by the evolution of the Lepidolina, a group of Permian foraminifera. Living during the middle and upper Permian throughout Southeast Asia as far north as Japan, the lepidolinas exhibit gradual variation in four out of nine features studied (Fig. 9.4). The largest change affects the diameter of the first chamber which shows an overall but irregular increase in size, apparently concurrent with the passage from shallow carbonate environments to deeper detrital sediments. This morphological change appears to have been controlled by environmental conditions which seem to have exerted certain selective pressures.

The Mole Rats – From Spain to Siberia. As a result of their high frequency in the palaeontological record and their abundance at individual sites, rodents are prime material for the study of gradual evolution.

The best documented example is that of the mole rats or water rats of Eurasia [8] which may be traced from the Pliocene species of southern France and nothern Spain to the present form, the water rat (Fig. 9.5). The lineage has evolved gradually over slightly more than 3.5 million years suffering extensive morphological changes. This evolution may be demonstrated by the analysis of tooth structure (variations in enamel thickness, and appearance of cement in the re-entrant angles) and by pronounced changes in tooth height which is easily visible. There is actually a gradual increase in tooth height, accompanied by the later appearance of the dental roots. There is also a general increase in size. In this lineage, the progression from the archaic to the current genus takes place through the eventual disappearance of the dental roots, the teeth themselves continuously increasing in size. The modern lineage is characterized by an increase in size, but even more so by a change in the thickness of the enamel in the dental cusps. The evolution of this lineage is very useful for dating prehistoric deposits [25].

The oldest part of the lineage has been analysed by statistical methods [25–27]. The results show that the lineage does not exhibit any trace of morphological sta-

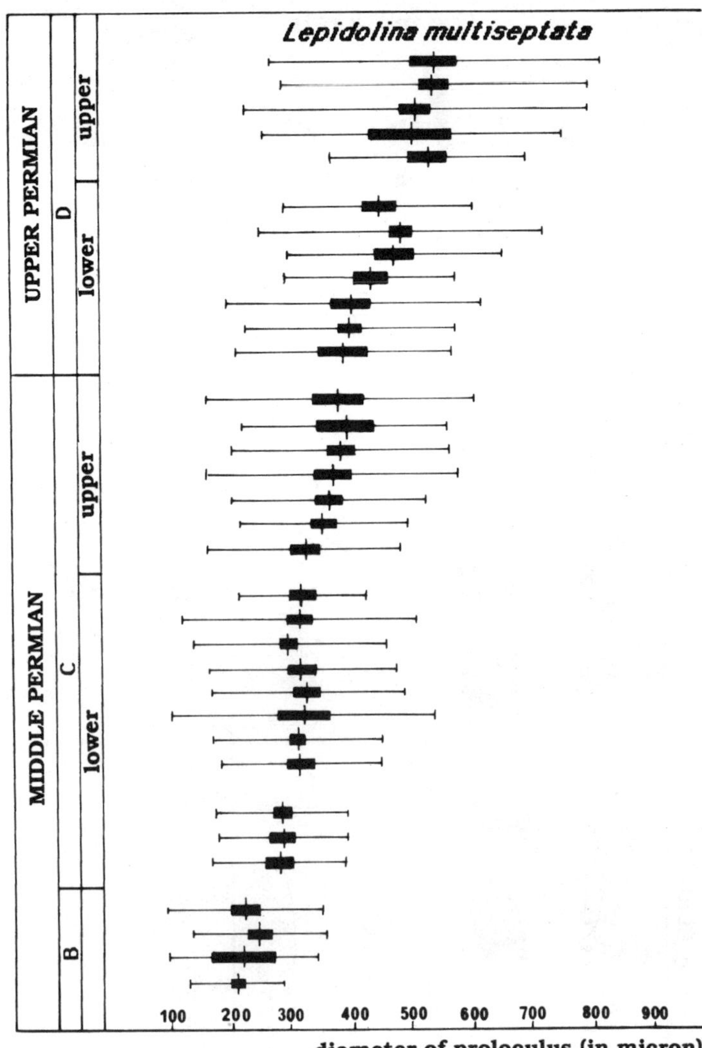

Fig.9.4. Evolution of Asian Foraminifer (Lepidolina). Permian foraminifera of the species *Lepidolina multiseptata* that extended from Southeast Asia to Japan, exhibit gradual variations in the size of chambers; this is directly influenced by fluctuations in the sedimentary environment. (After Ozawa [13].)

bility and that the morphological variation between successive populations shows a number of superimposed trends. As this lineage has been well dated by radiometric methods, pollen analyses and faunal associations, and by depositional events, it is possible to show that this gradual evolution took place at various rates and that rhythmic evolutionary changes are superimposed on changes resulting from climatic fluctuations. Climatic changes appear to have acted only as stimulants. The

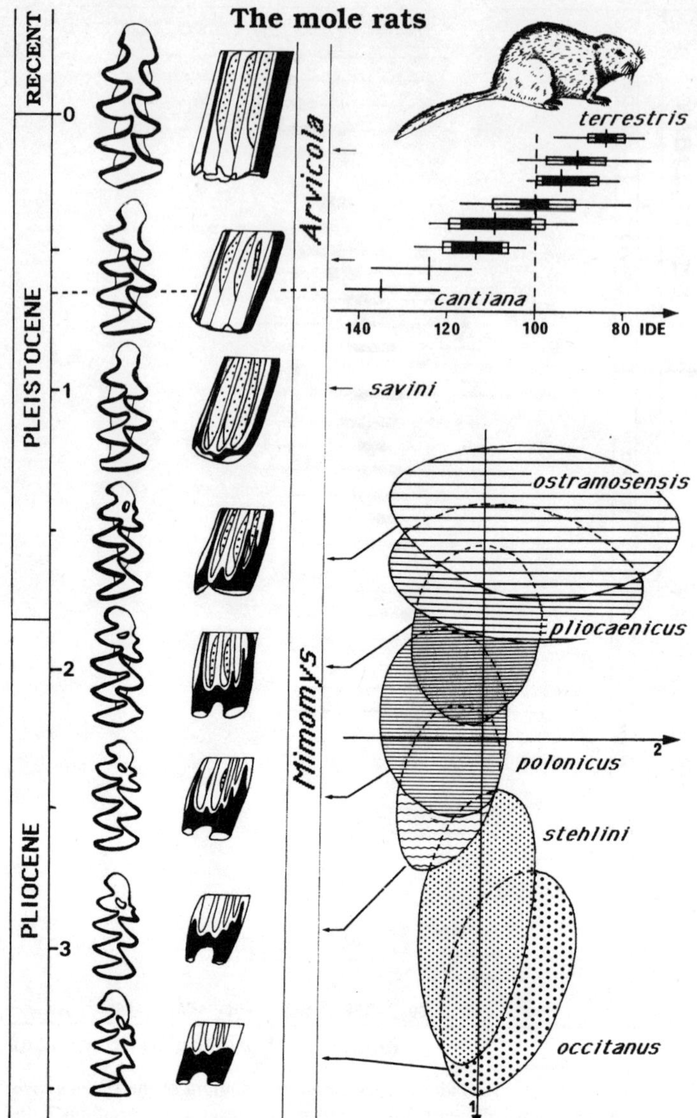

Fig.9.5. Gradual evolution of mole rats. The best example of gradual evolution in mammals is shown by the mole rats, rodents living in open fields. It has taken place on an almost global scale in Eurasia from Spain to Siberia from 3.5 million years ago to the present day. The phyletic gradualism covers increase in tooth height and various other morphological features. The analysis of morphological parameters shows a gradual overlap of successive populations. This phyletic gradualism is used for dating various Quaternary and prehistoric deposits throughout Eurasia. On the *left* the crowns are shown in top view and on the *right* in side view. Zones in which the enamel is interrupted are shown in *black,* the enamel itself is shown in *white,* and the cement in the re-entrant angles is *stippled.* The *right side of the figure* shows the overlap of successive populations. *The upper right* illustrates the gradual evolution of *Arvicola,* based on the thickness of enamel along the anterior and posterior sides of the dental triangles. (After Chaline [8]; Chaline and Laurin [26].)

morphological responses to similar climates may be highly variable (increase in dental height or appearance of cement) whereas responses to different climates may lead to identical results (increase in dental height). Other examples of gradual evolution are presented in Appendix 9.1 [28].

Repetitive Reversible Variations

As outlined in Chapter 7, there are reversible repetitive variations, referred to as ecophenotypic variations or phenotypic plasticity because they are controlled by variations in environmental factors. These variations appear to be especially frequent in snails [29–31] and nummulites, a type of fossil foraminifera [32].

The European snail (Pupilla pupilla). These are snails of the European Quaternary with a morphology directly indicative of environmental factors such as, in particular, temperature and humidity. In temperate regions the shells are low and short with a reduced number of whorls, and this contrasts with the higher, more elongate shells with considerably more whorls which are found in cooler regions.

The analysis of a large number of Quaternary from Achenheim (in the region of Alsace/France) revealed gradual morphological variations which are, however, repetitive and reversible, correlating with variations in climatic factors (Fig. 9.6). They do not represent gradual, cumulative, and irreversible variations as in phyletic gradualism, because they arise from an entirely different strategy of survival.

Stasis

A certain number of species have remained morphologically stable throughout their entire life. Examples are known from all groups, but their frequency appears to be higher among those groups exhibiting slow evolution. One of the classic examples for stasis is the small crustacean *Triops cancriformis* which has not changed morphologically since the Triassic, over a period of about 200 million years [9]. We should also mention those species referred to as living fossils, like the dipnoans, coelacanths, nautiluses, etc. [10]. Cases of stasis are also well documented among mammals such as some Tertiary herbivores of Wyoming (Fig. 9.7) [23].

It must be pointed out, however, that a morphological stasis observed in the palaeontological record does not necessarily imply that the lineage has remained stable at other organizational levels. Changes could have taken place at the genetic, ecological, or behavioural levels and these are not detected by palaeontology. The existence of such changes is suggested by examples in the living world, as shown for mice in Chapter 7.

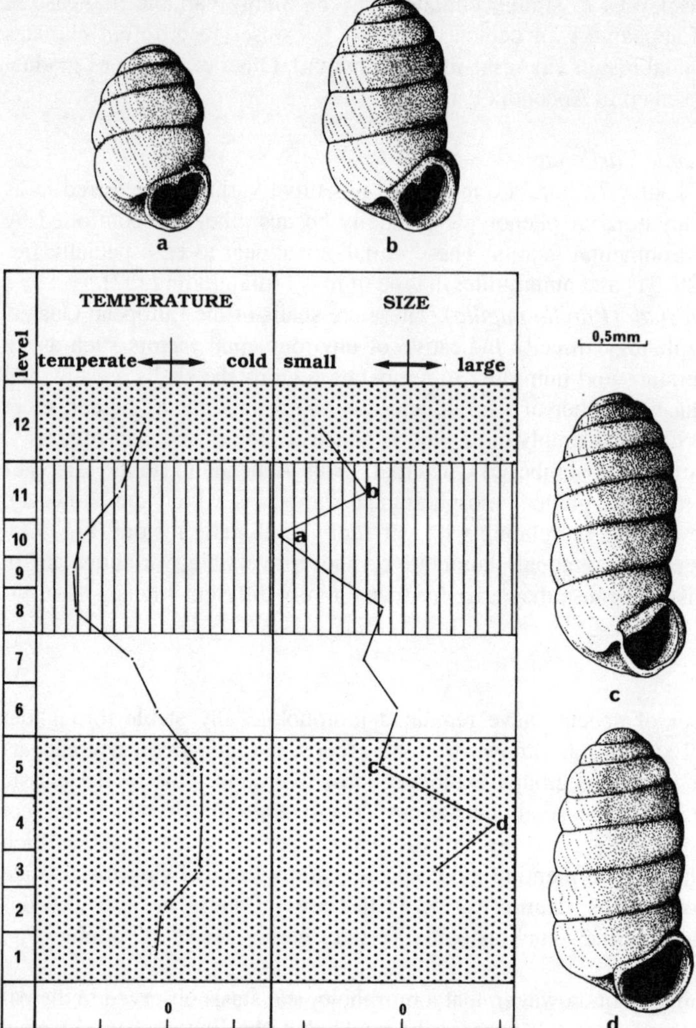

Fig.9.6. Reversible and repetitive evolution of the European *Pupilla,* a gastropod mollusc. The Quaternary loess profile of Achenheim (Alsace) has yielded vast populations of *Pupilla muscorum* in a well-defined chronological and climatic succession. The *lefthand* curve illustrates fluctuations in temperature of the depositional environment, with temperate climates to the left and cooler ones to the right. The section depicts, from bottom to top, loess horizons (1–5), clay loams (6–7), water-bearing levels of an interglacial palaeosol (Achenheim II; 8–11) and a second loess horizon (*12*). The palaeosol Achenheim II corresponds to a mid-Pleistocene interglacial which may be subdivided into two cooler phases. The size of the shells *(right)* changes gradually during the interglacial warming from small at the base during warm phases to larger during cooler phases. The size of the shells is thus an expression of the ambient climate, grading progressively from a cooler steppe environment to a temperate forest climate, and again to the more open steppe. The fluctuations in size entirely controlled by the climate, are usually referred to as ecophenotypic, reversible, and repetitive variations. The morphology of the shells varies between two extremes, stout (*a*) and elongate (*d*), connected by the intermediate forms *b* and *c*. (After Rousseau [29].)

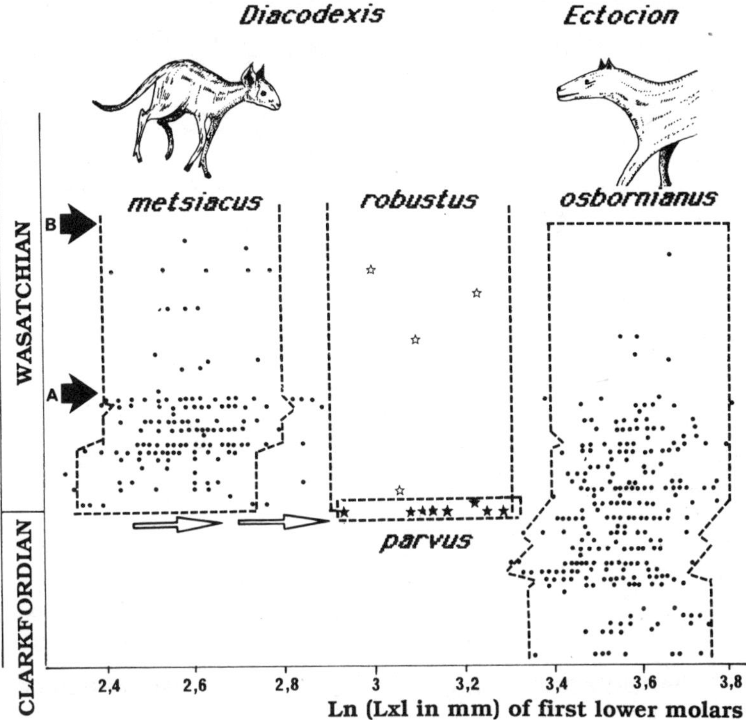

Fig.9.7. Examples of stasis among Tertiary artiodactyls. Of the Eocene artiodactyl lineages, *Diacodexis metsiacus, D. robustus,* and *Ectocion osbornianus* remained in morphological stasis, whereas other contemporary lineages evolved gradually (on the abscissa the log of the length and width of the first lower molars is plotted). (After Gingerich [23].)

A Mechanism of Evolutionary Change: Chronological Shifts in Development

Chronological shifts in development, so-called heterochronies, represent changes in the path of development from the egg through the foetal stage, birth, and youth, sometimes with phases of moult or metamorphosis, to the adult stage [35, 37–39]. Such changes may be the result of either retardation or acceleration of development, of the earlier or later appearance of sexual maturity which results in growth periods of different lengths, or of changes to the timing of the onset or completion of growth of specific structures.

The "Horned Wolves"
In order to explain the rather abstract concept of chronological shifts in development, a hypothetical animal, the "horned wolf", is considered in Figs. 9.8 [37].

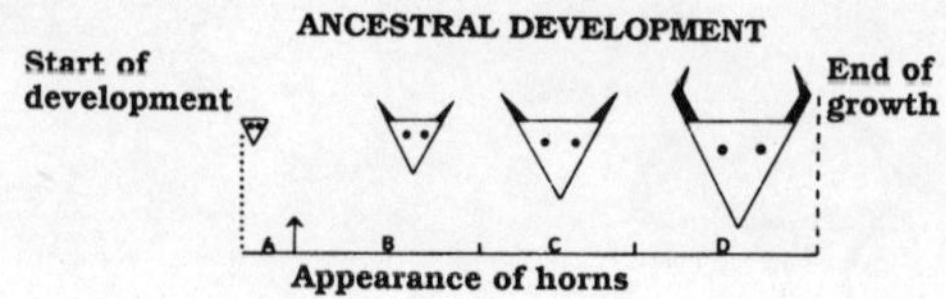

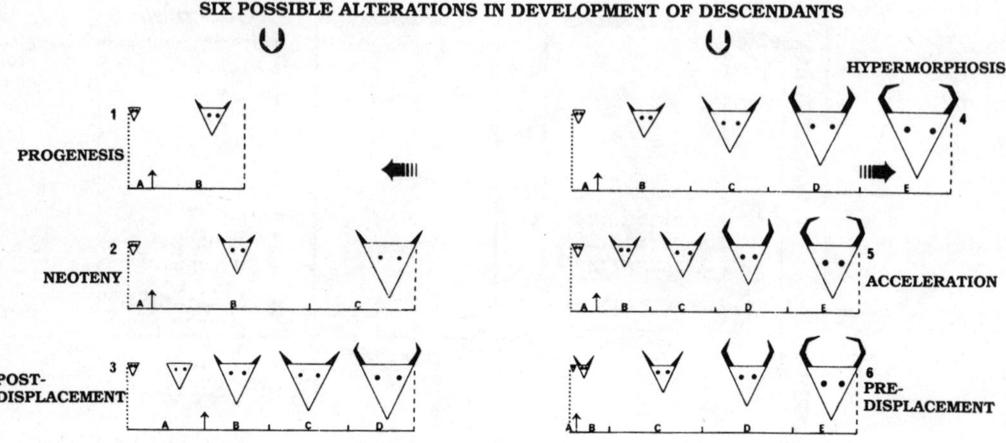

PAEDOMORPHOSIS **PERAMORPHOSIS**

Fig.9.8. The "horned wolves" or chronological shifts in development. *Above* The normal development of an individual of the imaginary species "horned wolf" (*A* first phase ending with the appearance of the horns; *B, C,* and *D* further phases of development. The end of phase *D* marks the end of growth after the onset of sexual maturity). *Below* The six changes or possible shifts in development. *Left* The process of paedomorphosis which permits the persistance of juvenile features into the adult stage. It may take place in the form of (*1*) progenesis, the termination of development through early onset of sexual maturity; through (*2*) neoteny, a slowing down of development which can affect all or part of the organism, but does not affect the size; or by (*3*) post-displacement during which the slowing down only affects a certain feature and not the organism as a whole. In post-displacement the onset of growth of a particular structure is delayed. The subsequent rate of growth is the same as in the ancestor, but because it starts later, it is not so advanced as in the adult. *Right* Processes of peramorphosis leading to the appearance of new hyper-adult features. These may result from (*4*) hypermorphosis, the extension of development due to the delayed onset of sexual maturity, or from (*5*) an accelerated rate of development. Another possibility is (*6*) pre-displacement when the onset of growth of a certain structure is initiated at a earlier growth stage than in the ancestor. (After Devillers et al. [37].)

The classic case of chronological change in development is neoteny. It results from the slowing-down of the development of an organism without affecting the organism's size. It leads to descendants which are of the same size as the original adult form, but which retain juvenile features. This process is observed in modern species which live in densely populated environments or which are exposed to strong competition with other species in a stable environment.

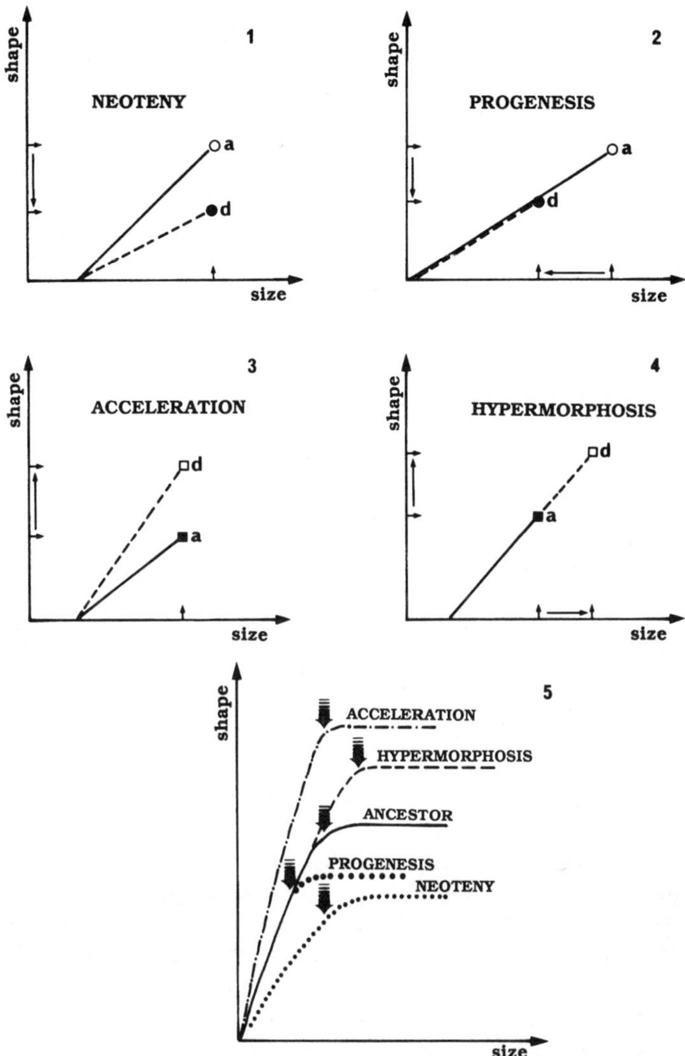

Fig.9.9. Patterns of heterochrony. Changes in the developmental pathway of an animal embryo or only of a certain organ. Size (x axis) is the largest dimension, and shape (y axis) the ratio of two dimensions. The *arrows* on the y axis marks the onset of sexual maturity, the beginning of the adult stage. *a* Normal (ancestral) development; *d* modified development. *1* Neoteny. Development of *d* is slowed down, although the time of onset of sexual maturity is not modified; the neotenic adult (*d*) attains the same size as the normal adult, but its shape is that of the juvenile stage of *a*. *2* Progenesis. The rate of development of *a* and *d* is identical, but sexual maturity appears earlier in the latter. The adult stage *d* is smaller than *a*, and retains juveline features. *3* Acceleration. *d* develops more rapidly than *a*, attaining sexual maturity at the same time as the latter; *d* will then acquire "hyperadult" features which did not have sufficient time to develop in *a*. *4* Hypermorphosis. The rates of development are identical, but maturity occurs later in *d;* the extended development period facilitates the formation of "hyperadult" features. *5* The four types of heterochrony. On each curve, the *arrow* marks the onset of sexual maturity. "*Ancestor*" signifies the normal development of a known or inferred ancestor. (*1–4* After Alberch et al. [38]; *5* after McNamara [39].)

There are numerous examples of this among the amphibians, the larval and adult stages of which are separated by metamorphosis (cf. Chap. 4; Appendix 4.2 the axolotl). In salamanders (newt) it has been observed that temperature is one of the factors affecting neoteny, which appears to be controlled by a simple hormonal process under the influence of two forms of the same gene. The environment thus appears to act as a stimulus initiating the chronological shift in development.

This is a very important mechanism for evolution. In the Triassic we observe, for example, that certain terrestrial organisms such as the temnospondylian amphibians returned to an aquatic life, leading to the birth of the sterospondylian amphibians. This transition from one group to another results in the partial or total elimination of metamorphosis which, in turn, leads to a slowing-down of ossification and to the persistance of cartilaginous tissues. Thus, more or less pronounced cases of neoteny represent a simple means for change between ecological niches.

When growth is slowed but there is still a constant rate of development, we are dealing with proportionate dwarfism. In the reverse case of the accelerated development of an organism, the adult morphology appears at an earlier stage. Size is not affected, and the resulting morphology is that of a "hyperadult" accentuated adult features.

As an example of acceleration consider the sea urchins *(Hagenowia),* for which an evolutonary tendency towards lengthening and reduction of the rostrum results from developmental acceleration.

When growth is accelerated at a constant rate of development, the end result is proportionate gigantism.

When sexual maturity appears at an early stage, further development of the organism is inhibited and the results is an adult of juvenile shape and size. We are then dealing with progenesis. This occurs in populations living in fluctuating environments where the population density is low and competition is minimal. This phenomenon is observed among salamanders of the genus *Bolitoglossa* which is made up of two terrestrial and one arboreal species. Salamanders of the latter species are smaller than those of the two former species; they resemble juvenile forms of the terrestrial species. The growth curve shows that the development is truncated, as is typical for progenesis. Here the chronological shift in development has permitted the occupation of a new ecological niche as the truncation of development supplied the adults with palmate legs [34].

From the Tertiary palaeontological record two unusual sea urchins have been identified with enlarged calcareous tests and teeth which were derived from forms in which the juveniles have teeth but the adults do not. Intermediate progenetic forms were very small and retained their teeth to the adult stage.

When sexual maturity is delayed development of the organism may continue over a longer time span. this extension of the growth phase facilitates the formation of a "hyperadult" morphology and a large size, a situation referred to as hypermorphosis. Examples of this are rather frequent in the palaeontological record. Among the captorhinomorph reptiles of the lower Permian, *Captorhinus* is small and possesses three rows of teeth in the upper jaw. However, forms derived during the upper Permian, such as the *Moradiscus*, grew very large, and at the same time, increased the number of teeth rows [35]. Another example is provided by the Irish Elk at the end of the last glaciation (Chap. 4).

A Global Species Concept

The information in Chapter 7 with a spatial and temporal approach permits the proposal of a global model of speciation that is characterized by occuring in a succession of phases [7, 8].

The changes taking place in isolates, leading to reproductive isolation, are referred to by biologists as "speciation *sensu strictu*". These changes actually define the establishment of a new species as the origin of new lineages of differing degrees of complexity and with differing life spans.

But will the changes occurring in the isolates cease after reproductive isolation has been achieved? Recent biological data casts doubt on this assumption. We have seen that reproductive isolation may be considered as a secondary effect of the modifications affecting small populations. We have pointed out that, for instance, with mice these genic or morphological modifications continue, depending on the circumstances of the expansion of a new lineage. They may result from mutations, irreversibly initiated processes, environmental selection, or competition with other species, etc.

Such phenomena take place rapidly on a geological time scale and do not have any statistical chance of becoming fossilized in the isolates. They therefore escape the notice of palaeontologists. Because of this, new species or lineages make their appearance rather abruptly in the palaeontological record; on the geological time scale they appear to be sudden events.

However, the changes initiated in the isolates, and especially those involving chronological shifts in development, will continue after reproductive isolation has been achieved. They take place from generation to generation and become discernible to the palaeontologist when a new species enters a phase of pronounced expansion. Palaeontologists have noted that the changes appear to follow two major tendencies.

The first tendency entails a morphological stability, also referred to as slow evolution or stasis. However, morphological stability does not imply stasis within all other levels of organization such as the genes, physiology, ecology, and behaviour. Changes at these levels may occur without the palaeontologist being able to detect any trace of them. A certain type of stasis involves genetic stability with morphological variation. The variations in shape are reversible as they are subject to fluctuations in environmental factors. This aspect has already been discussed in the case of changes in shape and size of continental and lacustrine snails as a result of fluctuations in temperature and humidity (phenotypic plasticity).

The succession of similar climatic fluctuations leads to the repetition of identical morphologies, referred to as repetitive and reversible variations. This flexibility of organisms permits the species to survive over vast areas subject to highly variable climatic conditions.

The second tendency, well documented in the palaeontological record, is gradual evolution. Among the changes taking place in the isolates and leading to the establishment of a new lineage, the beginning of a gradual irreversible morphological trend may often be observed and this then continues over several thousands or millions of years. These morphological changes during phases of gradual evolution are of such importance to palaeontologists that they consider them as a special type of speciation, the so-called temporal or phyletic formation of species.

In the spatial and temporal concept of species, morphological stasis, reversible and repetitive variations, and gradual evolution may be considered as delayed phases of speciation post-dating reproductive isolation. Consequently, a species is an independent entity, which is defined by a certain genetic programme and is in continuous evolution from its birth to its extinction.

This spatial and temporal concept of speciation occuring over time and in several phases before or after reproductive isolation may come as a surprise to biologists who are only rarely accustomed to the temporal dimension of a species (Fig. 9.10).

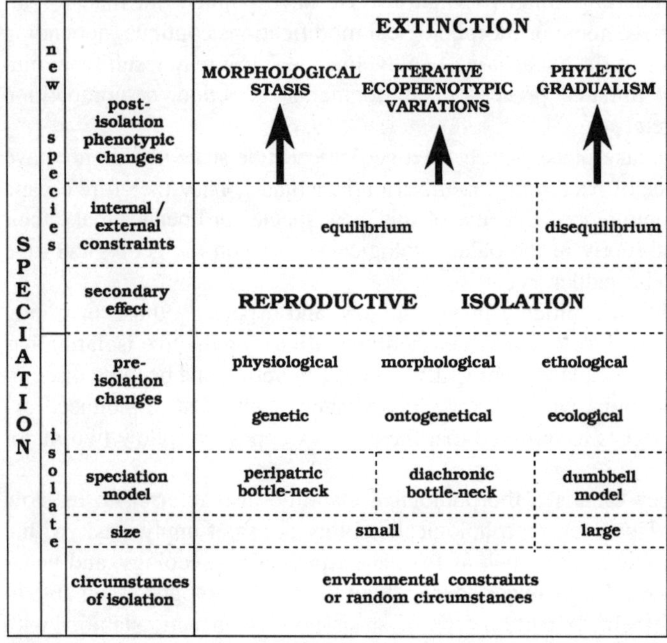

Fig.9.10. Temporal and spatial model of speciation. This integrates the temporal and spatial species concepts with traditional biological concepts of speciation sensu strictu and the phyletic speciation by palaeontologits. A new species frequently originates in an isolate from random circumstances or environmental constraints. As the number of individuals in these isolates is reduced, modifications may take place at any organizational level, from genetics to ethology of the organisms. As a secondary effect, these modifications lead to reproductive isolation from the source species, inhibiting any interbreeding between the two if the descendant species comes into sympatric contact with the parent species. However, under this new concept, the modifications will not cease with reproductive isolation. Palaeontological studies show that they continue until the eventual extinction of the new species. The modifications within a specific lineage take place in the form of reversible repetitive variations (ecophenotypic iterative variations) and by gradual evolution (phyletic gradualism), combined with morphological stasis. The latter, however, does not imply stability at other organizational levels of the species, for example, of a biochemical or chromosomal nature. (After Chaline [8].)

Integrated in the same spatial and temporal approache are "biological speciation" up to the appearance of reproductive isolation and "palaeontological speciation" taking place after this point. The different nature of these two approaches and their different scales of analysis, namely formation over geological time, had masked for a considerable time the "extent of speciation", which was envisaged as a dynamic phenomenon continuing throughout the existence of individual lineages from their origin to their extinction.

Biology and palaeontology, far from being opposed to each other in their objectives, are actually cooperating in the study of the global phenomenon of speciation. Biology contributes the mechanisms and processes, while palaeontology adds to this the historical aspects and the respective patterns. From this point if view, evolution may be rejoined with the study of speciation, as intuitively envisaged by Charles Darwin in *The Origin of Species*.

The evolution of lineages may result in considerable morphological changes. As we have seen, the action of chronological shifts together with other processes may explain the execution of the possibilities inherent in the genetic programme.

The history of life as revealed by palaeontology shows a succession of phases distinguished by the appearance of new features and by the multiplication of body plans and the number of groups now living. These groups then gradually succeeded in colonizing all ecological niches available on the continents and in the oceans.

How did these major body plans come about? What are the major trends in the evolution of life? We shall examine these questions in the next chapter.

Chapter 10

From Species to Body Plans

The key stages in the history of the living world may be summed up as follows: formation of the major types of body plan at the start of the Cambrian, with the appearance of the last, the vertebrates, during the upper Cambrian. Thereafter evolution took place within these plans through "modulations" exploiting a number of possibilities.

How did these major body plans come about? This is one of the crucial questions in evolution which we shall discuss in detail below.

How Did the Major Body Plans Form: By Micro- or Macroevolution?

For biologists like A. Vandel [1] and P.-P. Grassé [2], the processes of speciation would lead to divergences which are too small to explain the appearance of the major body plans of living organisms. Because of this, they proposed a distinction between a creative directed evolution, responsible for the formation of the various body plans and for new organs, and speciation proper, which assures diversifying evolution within these body plans. It is not a matter of a difference in the degree of change, but in its nature.

We would thus have to deal with macroevolution, a term introduced by R. Goldschmidt, which is responsible for the major body plans, and with microevolution, allowing the "dabbling" within these plans through the formation of species and their subdivisions.

Why should we require this concept, this distinction? The synthetic theory explains the appearance of new species and the major body plans by the gradual accumulation of small random mutations which are then subject to natural selection. Although this mechanism may be sufficient to explain the origin of species and their subdivisions, it appears insufficient for understanding the appearance of the major body plans.

This concept had already been advanced by R. Goldschmidt [3] and O. Schindewolf [4] who envisaged the appearance of the major structures as a result of particular processes like the "hopeful monsters", that is, by mutated individuals bearing new features. However, biological data show that major mutations are usually lethal for the respective individuals, and this thereby excludes a generalization of this mechanism.

Palaeontological data have revealed that the pronounced divergences presently separating the major body plans result, in part, from the temporal perspective. This is caused by the fact that as we go further back in time, the origin of the groups becomes blurred by the increasing number of gaps in the record.

Biological data show that the passage from one structure to another may either be gradual, through a variety of intermediate stages, or abrupt. The latter process prevents the appearance of any intermediate structure which would have had to be functional.

Among the important processes which seem to play a prominent role in the appearance of major structures, special emphasis must be placed on changes in development, like timing changes to the programme controlling the formation of an organism. We shall illustrate this with two examples.

Becoming Free? The origin of the Vertebrates

From which body plan could the vertebrates have developed? Virtually all "invertebrate" plans have been suggested as potential ancestors. We shall consider here only of these hypotheses.

The chordata, of which the "vertebrates" are a part, are assigned together with the echinoderms (sea urchins, starfish, etc.) to the same major group (Fig. 1.3). This would imply a common origin from an ancestor far removed in time. This remarkable association is justified by the fact that certain echinoderms (holothurians or sea cucumbers) and a small group of chordata possess rather similar larval forms. The argument is still wide open to discussion, but we shall restrict ourselves to looking for the potential origin of the vertebrates from within the chordata. Two of the groups (ascidians and *Amphioxus*) develop from a swimming larva which, by a metamorphosis, leads either to a sessile adult form (ascidians) or to a more or less sedentary form *(Amphioxus)*. According to one plausible hypothesis [5], a process of paedomorphosis sets in which eliminates metamorphosis to the adult stage and thereby preserves the free larval stage. From this the vertebrate construction developed which, among other features, is characterized by improved mobility and by the persistence of the free-living form. Together with the formation of a pronounced muscle system and loconmotory appendages around a supporting skeleton, arose the development of sensory organs for vision, touch, smell and hearing, as well as chemical stimuli, linked to a complex brain with which the animal could perceive the environment surrounding it, and at the same time react to such stimuli.

The Reptiles Which Turned into Mammals

The palaeontological record extending over more than 100 million years from the Carboniferous to the upper Triassic illustrates an evolutionary lineage referred to as "mammal-lile reptiles" which leads from "reptiles" through a number of modifications to the mammals. Let us discuss two of these changes which are functionally linked to each other [6].

Over most of this evolution, the lower jaw was made up of several bones connected to the quadrate bone of the skull by the articular bone (Fig. 10.1), as is seen in present-day reptiles. In the middle ear a skeletal element, the columella, transmits sounds from the tympanum to the inner ear.

In a mammal, the lower jaw is reduced to the dental bone connected to the squamosal of the skull. A chain of three small bones (malleus, incus, and stapes) transmit the sound vibrations through the middle ear which is enclosed in a bony capsule.

Embryology, anatomy, and palaeontology provided the data which show that we are dealing here with homologies between the malleus and the ancestral articular bone, between the incus and the quadrate bone, and between the stapes and the columella. These homologies prove that during evolution the articular bone was transformed into the malleus, the quadrate into the incus, and the columella into the stapes. In this way, a new articulation of the lower jaw was established without the animals concerned losing their hearing or their ability to feed.

Until only a few decades ago nobody could understand how this change had taken place. Through abrupt replacement? This would have represented a creation! Or through progressive change? This hypothesis appears improbable, but can be proven by palaeontological data.

The dental bone gradually extended backwards in length, eventually making up virtually the entire lower jaw, and thereby establishing the new type of articulations with the squamosal. At a certain point in time, the two types of articulations, reptilian and mammalian, must have been functioning side by side. *Diarthrognathus* [7] from the lower Jurassic of South Africa came close to this state. While the dental bone increased in size, the other bones became smaller and were reduced to a thin spike which, later still, lost its connection with the dental bone to eventually become incorporated in the small bones of the middle ear.

With this example we can illustrate in concrete form one of the major problems of any evolutonary research, that of evolutionary lineages. It is the task of palaeontologists to recognize and describe these. In the present case, the studies are well advanced. We then proceed to the second aspect of the problem, the question of "how"?, and "This requires the cooperation of biologists who study living organisms. Under what impulse is the "phenomenon of evolutionary lineages" initiated and what are the directive forces controlling this phenomenon over what are sometimes remarkable time spans? At this stage we are only able to offer some rudimentary answers to these questions. For this purpose we have to invoke some of the developmental mechanisms described above, and consider them in combination with natural selection. The latter process takes place in patterns which will force us to re-evaluate its true impact on organisms (Appendix 10.1).

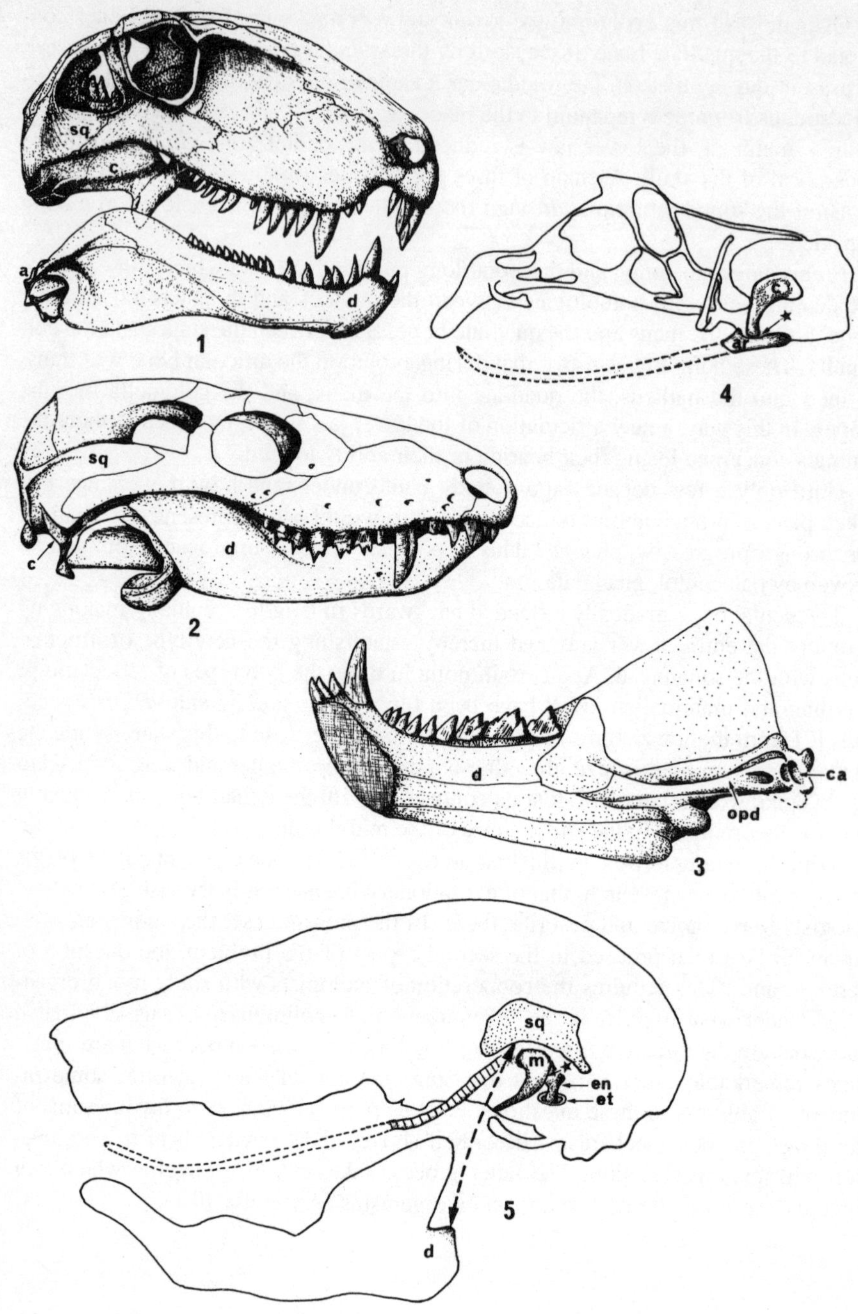
sq
c
a
d
1
sq
c
d
2
ar
4
d
ca
opd
3
sq
m
en
et
5
d

Chronological Shifts in Development: A Mechanism for Major Evolutionary Change

Chronological shifts in the timing of development (see Chapt. 9) actually facilitate the passage from an early structure to other types of structures without considerable genetic changes. All that is necessary is modification of the regulatory or temporizing genes which control the timing of the expression of structural genes.

This is a flexible mechanism, permitting a certain structure to evade its developmental constraints and to become another structure built under new constraints. With this mechanism, nature is able to exploit the potential included in the various genetic programmes and to execute a number of possibilities through deviations in development.

It is conceivable that chronological shifts in development, along with jumping genes or transposons, could constitute one of the major processes forming new structures and this would also explain rapid changes in genetic programmes.

Thus there is no justification for a distinction between micro- and macroevolution, or for considering separately the results of evolution at the species level from those which are amplified with time by the accumulation of speciations, to appear at the higher organizational levels of the genus, family, order, class, or phylum.

One of the arguments advanced by the evolutionist James W. Valentine [8] in support of macroevolution is the fact that most of the major groups appeared during the Cambrian 500–570 million years ago. If the processes of speciation, microevolution, alone had been active, then all living organisms would have one basic body plan as, according to this author, microevolution only works by a slow and progressive accumulation of small mutations and would therefore be unable to give rise to complex body plans.

This was the traditional concept of the synthetic theory which was based on a restricted knowledge of genetics, and did not take into account the extent of structural modifications resulting from chronological shifts in development.

Fig.10.1. The reptiles which became mammals. *1 Dimetrodon,* mammal-lile reptile of the Permian pelycosaur group. The dentary bone *(d)* comprises two-thirds of the mandible. The very large quadrate bone is fixed to the skull and the articulation of the jaw is effected between the quadrate *(c)* and the articular bone *(a)*, a small bone at the back of the mandible. (*sq* Os squamosum.) *2* In the Triassic *Trinaxodon,* the quadrate bone *(c)* has become a small bone unconnected with the skull, and the dentary bone *(d)* makes up almost the entire jaw *Thrinaxodon. 3* The lower jaw of the Triassic *Cynognathus,* as seen from the inside, shows the reduction of the group of postdentary bones *(opd)* to a small rod fixed to the dentary bone *(d)* which makes up almost the entire mandible (*ca* Articular cavity on the articular bone). *4* and *5* Cartilaginous skulls of lizard *(4)* and hare *(5)* embryos. The outline of the brain case is only indicated, emphasis being placed on the lower jaw and its articulation with the skull. In *5,* the major portion of the cartilaginous jaw has already degenerated *(stippled);* the *hatched portion* will eventually also disappear. The completely developed dentary bone has been moved downwards to expose the articular zone of the cartilaginous jaw. In these two sketches ★ marks the same articulation, articular-quadrate for the lizard and malleus *(m)*-incus *(en)* for the hare (*et* stapes*). The placement of the final articulation in the hare is indicated by the arrow* between the dentary bone *(d)* and the squamosum *(sq)*. (*1* After Romer [17]; *2* after Parrington [18]; *3* after Crompton [7]; *4* after de Beer [19]; *5* from the authors.)

Evolution Within the Major Body Plans

Gradual evolution is an evolutionary mode active within a lineage and should not be confused with so-called evolutionary trends.

Evolutionaly Trends: The Sabre-Toothed Cats
In palaeontological terminology, evolutionary trends represent a type of evolution during which one or more features are always modified in the same direction. An evolutionary trend may be maintained by several successive species which reinforce the feature in question. There may be variable rhythms within an evolutionary trend, no doubt in combination with phases of stabilization between the speciations.

An excellent example of an evolutionary trend [9] is provided by the sabre-toothed cats, the so-called machairodont carnivores (Fig. 10.2). The trend is characterized in particular by increasingly extended growth of the upper canines which reach, almost like daggers, beyond the lower edge of the lower jaw. This downward extension of the upper canines is accompanied by parallel modifications of the lower part of the lower jaw. These modifications fulfill a protective function against breakage of the downwardly protruding teeth. Along with this trend there is an increasing degree of reduction of certain teeth in the rear part of the jaw, and the skull shape becomes modified under the functional constraints of mastication. This array of orientied changes is accompanied by an overall increase in size in these animals.

Another classic example of evolutionary trends is presented by the titanotheres, a group of rhinoceros – like mammals which are characterized by an increase in body size and in the size of the horn on their heads (Chapt. 4).

Are the Trends the Result of Interspecies Selection?
According to S. J. Gould and N. Eldredge [10], evolutionary trends are the product of differences in the success of the species during competition in nature after the diversification of a group. This concept is also subscribed to by S. M. Stanley [11] who proposed a particular type of selection as acting between species. According to this, the trends result from the survival of certain species after biological competition has eliminated the less successful species.

With this we actually enter another organizational level of the species, namely strong and complex communities. This requires further investigation because interspecies selection represents a concept which has been little studied so far.

Evidence From the Gnus (Wildebeest). An interesting example present by Elisabeth Vrba [12], is that of the African antelopes (Fig. 10.3). According to this author, the trend of diversification of the hartebeests and gnus is only a secondary effect of the rate of multiplication and extinction which, in turns, is related to the degree of specialization of the respective species to their environment. The so-called non-specialist species are able to exist in different environments, feeding on a variety of vegetation. The specialist species, however, live in more restricted niches and thus are less widely distributed. There actually appears to be a connec-

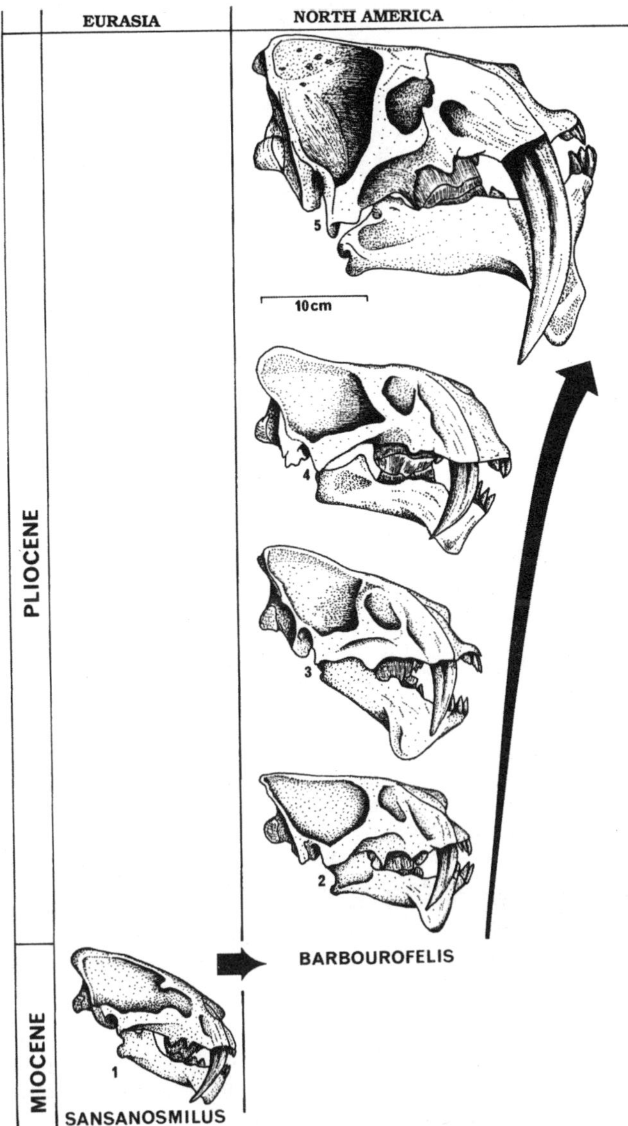

Fig.10.2. The sabre-toothed cats, an example of evolutionary trends. Making their first appearance in Eurasia and Africa during the Miocene, the sabre-toothed cats of the "Barbourofeline" group had migrated to North America by the upper Miocene. The lineage evolved by increasing in size, representing an excellent example of an evolutionary trend in the form of a remarkable lengthening of the upper canines. This evolution entails other morphological modifications from a functional point of view. The lower jaw had to acquire a wider arc of rotation facilitated by the vertical arrangement of the occipital zone. The articulation of the lower jaw was shifted lower, the upper carnassial teeth became enlarged and the premolars reduced. The Barbourofelines were widely distributed throughout Eurasia (from France to Mongolia) and Africa during the Miocene. A northern Asiatic population must have migrated to North America about 11 million years ago. *1 Sansanosmilus palmidens; 2 Barbourofelis whitfordi; 3 Barbourofelis morrisi; 4 Barbourofelis lovei; 5 Barbourofelis fricki*. (After Martin [9].)

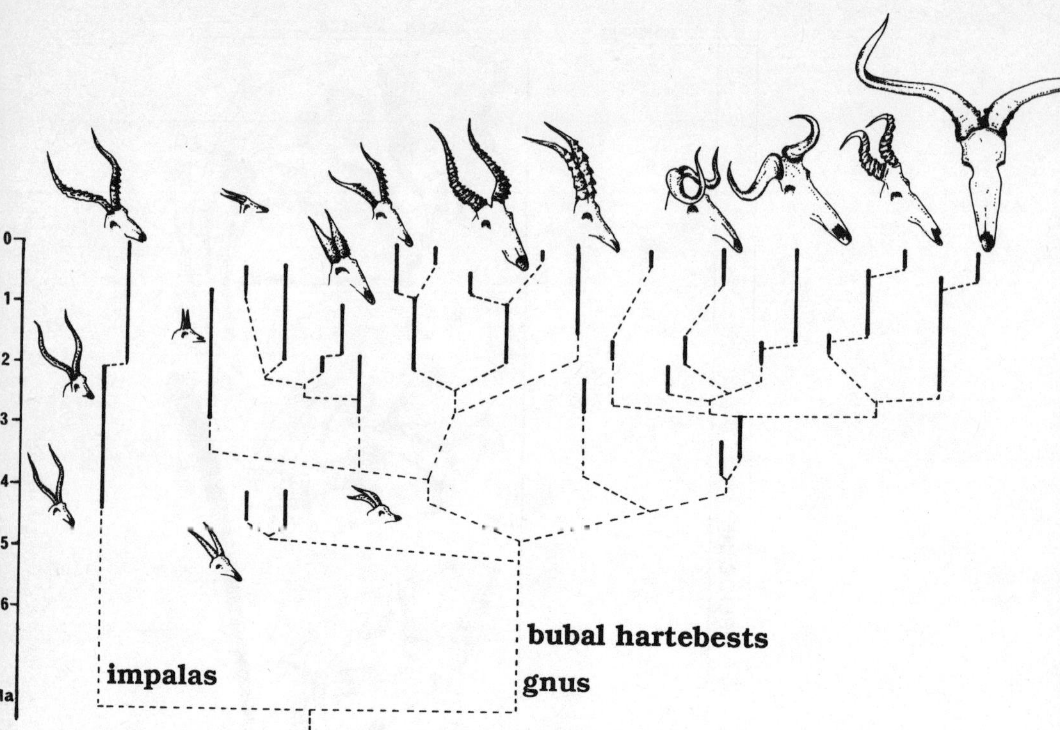

Fig.10.3. Radiation of African antelopes. This phyletic tree illustrates the parental relationships between the impalas (Aepycerotini) and the gnus and hartebeests (Alcelaphini). As non-specialized feeders (generalists), the impalas can live in a variety of environments and, as a result, they have experienced fewer specific diversifications than the gnus and hartebeests. The latter two groups have more imperative alimentary constraints (specialists) and, as a consequence, the various species occupy a large number of more restricted ecological niches, leading to a wide range of speciations. This difference in diversification between the two groups is not the recent expression of a progressive selection with differential success of the species. Actually, in most fossiliferous Plio-Pleistocene strata, the impalas are represented by one abundant species only, whereas the gnus and hartebeests are represented by seven lineages. (After Vrba [12].)

tion between the morphological diversification of an organizational type and the utilization of available ecological niches for the specialization of the species. The evolutionary trends could result from the exploitation of certain factors inherent in the genetic programme when the occupation of neighbouring niches leads to similar specilizations. Increasing trends of climatic changes over geological time could then favour the strengthening of certain adaptations. Supporting evidence for this assumption has yet to be demonstrated.

The Early Tertiary Mammals Communities of Quercy. Research by Serge Legendre [13] on early Tertiary mammal communities studied by an original ecological method (Fig. 10.4) has shown that these communities exhibited constant structures over long periods when the distribution of the species was controlled by their numbers. Stability was maintained even when the global evolution of the respective lineage led to an increase in size. This implies that in order to compensate for the global trend to a larger size, the larger species became extinct and were re-

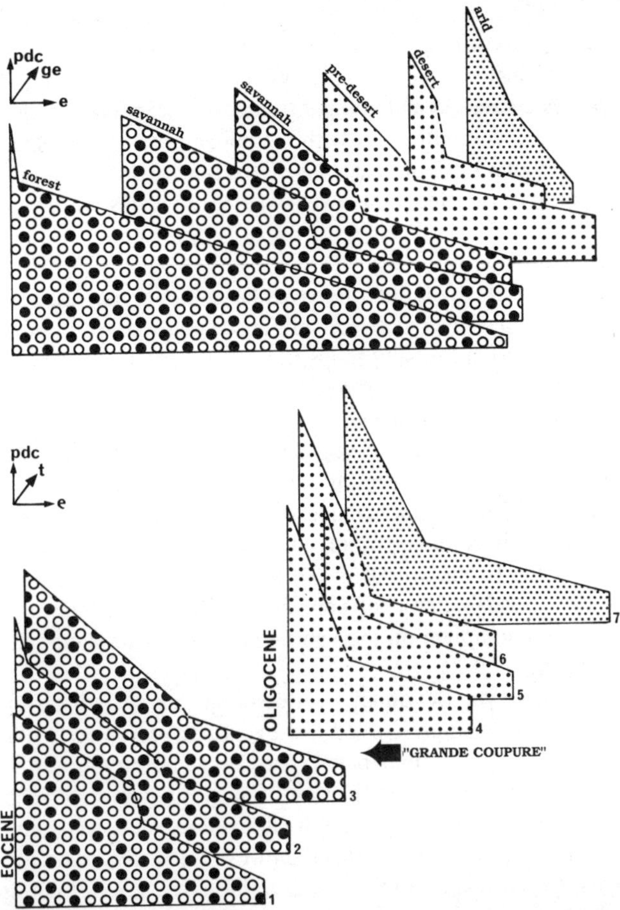

Fig.10.4. The early Tertiary mammal communities of Quercy. *Above* for comparative purposes cenograms (or ecological diagrams) of recent mammal communities are shown for forest (tropical rain forests of Gabon), savannah (tropical tree savannah of Rwanda and Zaire), pre-desert (northern Sahara of Algeria), desert (Iran), and arid zones (open arid mediterranean environment of Spain). They are based on the distribution of the species weights *(pdc)* and the number of species *(e)*; a certain ecological gradient is shown by *ge*. In savannah and desert, there is a lack of medium-weight species whereas in the tropical forest, the distribution of weights is more even. *Below* an analysis of faunas at the Eocene-Oligocene boundary. The pronounced change in the faunas, "*major break*", marks the passage from the tropical rain forest or tree savannah to desert or arid regions. *1* Le Bretou; *2* Perrière; *3* Escamps; *4* Aubrelong I; *5* Mas de Got; *6* Pech Crabit; *7* Pech du Fraysse; *t* time. (After Legendre [13].)

placed by smaller animals; thus there was an equilibrium between the appearance and disappearance of species. Some specialists have suggested that the probability of becoming extinct increases with the size of the species, as the size increase of mammals correlates with a decrease in population density and thus in the number of individuals.

Palaeontological data on the famous Quercy faunas show that at the end of the Eocene (40 million years ago), mammals underwent a phase of extinctions caused by a change in the climate. As a result of this disequilibrium, the number of species decreased and competition between species became less pronounced. This led to a new equilibrium with the appearance of new species during the mid-Tertiary about 33 million years ago. This type of study will undoubtedly allow the real importance of selection between species to be tested.

Radiations
Studies of the evolution of groups have shown that once a new body plan has been created, rapid diversification will ensue and will result in numerous lineages which themselves lead to subgroups. Such adaptive radiations are the expression of exploitation of a number of possibilities within a given body plan by the pronounced multiplication of lineages.

Ammonites. The diversification of the ammonites is a well-documented example which has recently been reassessed by H. Tintant [14]. Ammonites are marine molluscs which made their first appearance during the lower Devonian, eventually becoming extinct along with the dinosaurs and some other groups at the end of the Cretaceous (Fig. 10.5). Their origin must be found within the orthoeratids, another group of cephalopod molluscs which possess a straight external shell with a siphon on the ventral side and often a curved suture line partitioning the shell.

In the lower Devonian a series of species was discovered, the shells of which became increasingly curved and eventually completely coiled. This coiling of the shell gives the animal better buoyancy, improving its vertical and lateral manoeuvrability. The modification of shape is generally accompanied by a crenulation of the suture line, a major innovation, although some heteromorphs without ornamentation have crenulate sutures; e.g. *Eubaculites* of the lower Cretaceous. This diversification took several million years, less than one geological substage, to become established. It represents a radiation during which all ammonite features made their appearance. During the remainder of their 300 million years of evolutionary history, the ammonites did not develop any new features as the "ammonite" body plan, the basic design or software of the group as it is termed by H. Tintant, had been established.

The ammonites then underwent a very complex diversification whith about 1000 genera and several tens of thousands of species. Their evolutionary history may be subdivided into three phases, each terminated by a crisis, namely, at the end of the Permian, Triassic, and Cretaceous.

The first radiation occurred during the Devonian when the goniatite group began to exploit the "ammonite" design by elaborating complications of the suture line of septa (appearance of new lobes and subdivision of folds), and by forming tubercles and ribs.

Regression of the shallow seas at the end of the Permian almost led to the complete disappearance of the group and only one lineage survived. This experienced a new radiation, leading to the Triassic ceratites which possessed septa that were more folded than those of the goniatites.

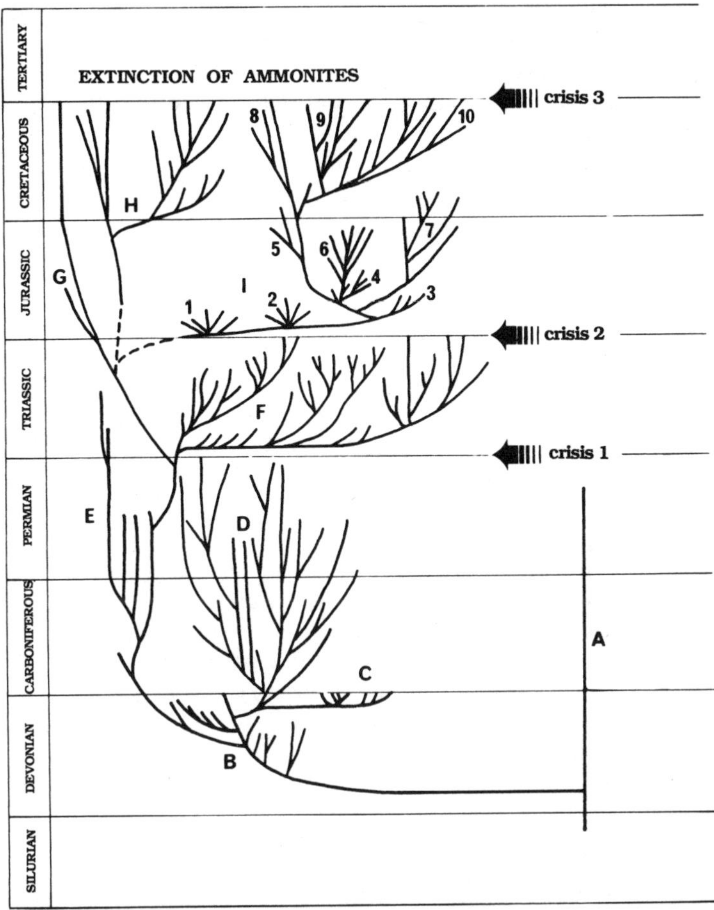

Fig.10.5. Ammonite radiations. Starting from straight forms like the Bactrites *(A)*, the first radiations gave rise during the Devonian to the Anarcestines *(B)* and the Clymenias *(C)*, and during the Carboniferous to the Goniatites *(D)* and the Prolecanitites *(E)*. The first major crisis *(crisis 1)* took place at the end of the Permian, and was followed by the Triassic radiation of the Ceratites *(F)*. The second crisis *(crisis 2)* at the end of the Triassic was followed by the radiation of the Phylloceratites *(G)*, the Lytoceratites *(H)*, and the Ammonitinae *(I)*. *Crisis 3* at the end of the Cretaceous led to the complete extinction of the ammonites. (After Tintant [14].)

A second crisis at the end of the Triassic eliminated the ceratites except for one lineage which experienced a new radiation, leading to the ammonites. These became extinct at the end of the Cretaceous.

During the succession of the three ammonite radiations a number of features (changes in suture lines, shape, ornamentation) were repeated and within each of the radiations, during the Triassic, Jurassic, and Cretaceous, the shells started to unroll again. In this case Evolution was quite repetitive, although it did not lead to identical shapes.

The alternation of radiations and extinctions observed among the ammonites appears to be the fate of major body plans in the living world. the phases of radiation frequently take place rather rapidly on a geological time scale and new groups seem to become established quite suddenly. We refer to this situation as the accidental aspect of radiation. Thereafter, the exploitation of the various organizational subplans leads to a considerable number of speciations, the results of which then occupy the available ecological niches. Rates of evolution differ between species, with evolution being gradual or punctuated. Eventually, the phases of partial or total extinction, also of an accidental nature, are related to global environmental crises.

The evolution of the ammonites permits us to observe the key role played by chronological shifts of developmental pathways in the diversification of lineages [15]. It is possible to observe these changes as the shells retain all the growth stages from juvenile to adult during their progressive size increase.

The ammonites are thus a prime group in which to study evolutionary patterns. Regrettably, their extinction has robbed us of the possibility of investigating their genetics, biochemical constitution, the significance of the numerous forms and the mechanisms of the speciations.

Studies of the nautilus which, like the ammonites, was derived from orthocerian molluscs, and which is represented by two species living in the equatorial Pacific, may allow us to narrow the gaps in our understanding of these animals. H. Tintant [16] has shown that the evolution of the nautiluses has been greatly underestimated. This resulted from the preconceived idea of their status as living fossils and from the multiple repetitive evolution complicating their systematics and our understanding of their evolution.

Chapter 11

Is Evolution Continuous or Discontinuous?

Extending the species concept in time leads us to understand evolution as a dynamic process which takes place at different rates and in different patterns. We shall examine this below.

We have already shown that the appearance of the major body plans does not necessitate any particular "macroevolutionary" processes which would differ from the mechanisms of speciation or microevolution.

Do Species Form Gradually or by Abrupt Discontinuities?

This old question has not lost any of its importance as concepts of continuity and discontinuity in evolution have succeded or opposed each other in the past [1].

Plato's concept of essentialism claimed that the types remained stable, thereby negating any idea of evolution. This approach was maintained until the end of the 18th century in the concept of uniformitarianism which envisaged the world as an eternal repetition without any history.

J. B. Lamarck was the first to develop a concept of transformation of the living world: living organisms form a continuous assembly, a veritable scale of complexity, which individuals enter or leave according to their needs. Lamarck thus subscribed to a so-called nominalist species concept which is based solely on the individuals arranged in a scale of species which is created by the human mind, but does not possess any objective reality. This theory was opposed by the creationist concept derived from the Bible, according to which species were created separately after having been destroyed in the Deluge. The discovery of different successive faunas in the fossil record necessitated a multiplication of catastrophes, deluges and discontinuities, leading to the concept of catastrophism, the a series of "revolutions" of the globe, developed by G. Cuvier [2] and Alcide d'Orbigny.

By reconsidering the transformist ideas of J. B. Lamarck, C. Darwin was the first to present the idea of evolution on the basis of scientific reasoning. This introduced to contemporary philosophical concepts an important approach, namely gradual change (or gradualism) which was to become a veritable dogma. This idea was widely accepted at the end of the 19th century, especially in England where there was a general hope for the continuous improvement of human life through the advance of science and its industrial applications. Repeating an expression ascribed to C. Linné "Nature does not perform jumps", this concept became gradual evolution (or phyletic gradualism) within the history of evolution.

Based on the discontinuities between species observed in the fossil record and on the presence of species which have been stable over millions of years (stasis), S. J. Gould and N. Eldredge proposed a stepwise concept of evolution in their model of punctuated equilibrium [3–4]. According to these authors, species are formed rapidly and all morphological changes are concentrated in periods which are short on a geological time scale. Thereafter, species remain stable, or in stasis, over their entire life spans. Although this model does not categorically reject gradual evolution or phyletic gradualism associated with minor adaptive modifications, it practically refuses to admid its existence, reducing its role to virtually nil. This brings us back to a view of discontinuous evolution, based partly on a structural (typological) concept of the species.

Palaeontological and biological discoveries have now given way to interpretations of continuous or discontinuous speciation. The various concepts rest on data which are obtained from observations on isolated cases and extended to include the entire living world. The apparent contradictions between the theories result from these rash genereralizations which are also extended to different organizational levels within the living world (genic, morphological, etc.). Recent data permit us to address this problem more broadly.

The Gradual Model of the Synthetic Theory: A Revision of Concepts

When we are talking about evolutionary patterns we try to define the ways and circumstances under which evolutionary phenomena take place. In the synthetic theory of evolution, G. G. Simpson {5–6] and Julian Huxley [7] postulated that evolution is gradual, although they admitted that there are also phases of very rapid evolution (quantum evolution). Evolution results from the accumulation of small mutations under the control of natural selection; from a genetic point of view this is caused by progressive changes in gene frequency.

In the synthetic theory, two evolutionary patterns are recognized: anagenesis and cladogenesis (Fig. 11.1). Anagenesis corresponds to evolutionary progress within a lineage, whereas cladogenesis ensures the multiplication of lineages by splitting a source species into two or more descendent species.

Gradual Evolution of Lineages

Anagenesis describes the establishment of a unique lineage and assures the continuation of a species. It may take place by different patterns. In most cases, anagenesis translates into the gradual irreversible evolution observed in fossil lineages. Palaeontologists frequently refer to this as "phyletic formation of species" because of the important morphological changes observed. In Chapter 9 we incorporated this as a late phase of speciation. For many palaeontologists, anagenesis

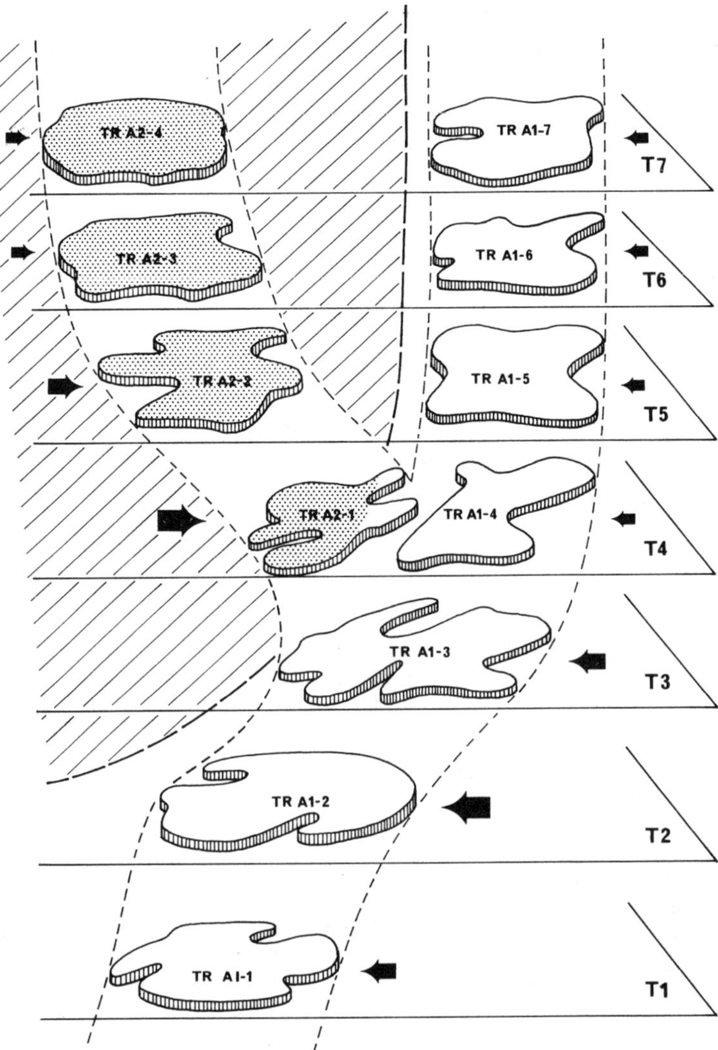

Fig.11.1. Gradual model of the synthetic theory. The synthetic theory is based on a model of gradual evolution with accumulating mutations and changing gene frequencies under the direct influence of natural selection. Adaptation is seen as the main driving force of evolution. The illustration shows the gradual evolution of a lineage (species A) represented at times *T1–T7* by the transients *TR A1-1* to *TR A1-7* (populations in temporary transition or stages of evolution within a lineage). At time *T4*, a random event (geographical barrier, climatic change, appearance of a new ecological niche) caused species A2 to divide into two groups, one of which (A2) started to colonize a new ecological niche or a new adaptive zone. By gradual divergence, lineage A2 started to evolve from *T4* to *T7* through the transients *TR A2-1* to *TR A2-4*. The evolution of lineage A1 at times *T1* to *T3* represents an evolutionary pattern referred to as "anagenesis", namely, evolution within a lineage. At times *T4* to *T5*, the splitting of lineage A2 from the source lineage (A1) corresponds to "cladogenesis", a division into two lineages. The size of the *black arrows* illustrates the influence of natural selection, with the environment playing a major role in evolution. The *hatched area* shows the new adaptive zone. (After Chaline [19].)

has become a derogatory synonym for gradual evolution. Thus the synthetic theory does not include under anagenesis cases of stasis or reversible repetitive variations which prove a certain stability of the species. However, anagenetic mechanisms should include speciations resulting from a temporary reduction in individuals which ensures the survival of a lineage but not its multiplications. In this case, a descendent species replaces a parent species in continuity as a result of a crisis in the number of individuals; "palaeontologically" this gives the impression of a discontinuity.

Analyses of speciations show that the subdivision of a lineage into two or more descendent lineages (cladogenesis) will actually result in complex patterns of speciation which, by themselves, do not correspond to particular circumstances or mechanisms of evolution. We are dealing here with the circumstances of speciation described in Chapter 7. When the isolates contain large numbers of individuals, new species will establish themselves through the patterns, processes, and mechanisms characteristic of the dumb-bell modell. When the numbers of individuals in the isolates are small, species will develop according to the model of spatial restriction of peripheral populations (bottleneck).

Cladogenesis and anagenesis appear to result from differences in the patterns of speciation. They should no longer be considered as particular patterns of evolution, but rather as a "secondary effect" of the phenomena of speciation. This represents a new approach which contrasts markedly with the synthetic theory of the 1940s. The evolutionary patterns now actually correspond to the different circumstances under which new species are forming.

The Punctuated Equilibrium Model

Publication of the punctuated equilibrium, or punctuational model by N. Eldredge and S. J. Gould has led to a fervent international debate on the patterns and rhythms of evolution. We shall try to analyse its content and implications before embarking in a critical discussion.

In the first version of 1972, the punctuational model was presented as an "alternative" to gradual evolution, as a replacement for the "gradual dogmatic" interpretation of the palaeontological record by the synthetic theory. It is true that palaeontologists have tacitly accepted the synthetic theory and that authors of evolutionary studies consider the palaeontological record exclusively from a gradual point of view, under which gradual evolution may advance at a slow, medium, or rapid pace. No palaeontologist would want to doubt this dogma. Most specialists would be content with describing new species, performing applied palaeontology, or rendering a service to geologists by giving them information on the chronology of the terrain under investigation. They would prefer to leave it to the "great thinkers" to mull over evolutionary problems.

In the eyes of palaeontologists, S. J. Gould and N. Eldredge were iconiclasts attacking the synthetic theory. These authors started from two basic observations on the palaeontological record of intervertebrates, in particular trilobites and molluscs. The first of these observations relates to the frequent existence of species which were stable for millions of years, while the second deals with the transformation of one species into another in the palaeontological record. When an ancestral species is found at a low stratigraphic level and a descendent species is discovered at a higher level, the intervening levels frequently do not yield the intermediate forms expected from the concept of gradual evolution. Moreover, the descendent species sometimes exhibits pronounced morphological differences from the parent species.

In the framework of the synthetic theory, these discontinuities were explained by imperfections in the palaeontological record, and it was argued that the gradual transition would be observable in other areas with more complete sedimentation. For S. J. Gould and N. Eldredge these discontinuities "cannot be ascribed to imperfections in the palaeontological record". As an explanation, they offered the "punctuated equilibrium model" which is based on the processes of speciation by geographical isolation accompanied by a reduction in the number of individuals, as described by E. Mayr (Chapt. 7).

The species are very stable groups remaining "in stasis" throughout their life or, in short, "stasis is data". These phases of stasis had been recognized by G. G. Simpson as forms of slow or bradytelic evolution. However, they had, not been considered of further value as they did not explain the changes, the sole objects of study. Thus, species remain in long phases of stasis in equilibrium with their environment. This stability, however, is interrupted or punctuated by rare events of speciation according to these authors.

The punctuated equilibria assert that the abrupt appearances (of new species in the palaeontological record) represent the expected expression over geological time of the conventional processes of speciation. Such processes take long compared to our human time scale, but are usually (although not always) instantaneous by geological time scales. Stasis is the logical expression of a good state of adaptation of large populations [4].

The daughter species develop marginally in a neighbouring territory. After having diverged sufficiently to acquire some degree of reproductive isolation, which permits them to come into contact again with the mother species without mixing, they are generally more competitive than the latter. As a result, they will eliminate the mother species from their original territory and occupy their place. This explains why in the palaeontological record ancestral species are abruptly replaced by their descendent daughter species.

For S. J. Gould [8], this model had two implications for evolutionary rhythms and patterns. "Evolution is neither slow nor regular, but the result of rare and complex events which rock the stability of the species. I would like to introduce for this the expression "punctuated equilibria". "The conventional concept of two evolutionary patterns, viz anagenesis allowing progress and cladogenesis assuring diversity is wrong. Real anagenesis is rare and of little importance. On the contrary, the trends are the products of the differential success of the species. The fundamental event in evolution thus is speciation. At a higher level, the trends are the result of numerous cumulative episodes of speciation." The latter point has been analysed in the preceding chapter.

From this comment by S. J. Gould we may retain the key role played by speciation and the minor role of anagenesis, which corresponds to gradual evolution. In 1977 the authors insisted:

"Our model of punctuated equilibria is a hypothesis about mode. We claim that speciation is orders of magnitude more important than phyletic evolution as a mode of evolutionary change... We do regard punctuated equilibria as by far the most common tempo of evolution – and we do assert that gradualism is both rare and unable in any case – to serve as the source for major evolutionary events."

In the same article, the authors methodically dismantled some classic examples of phyletic gradualism by showing that these had been poorly documented and that they could be better interpreted in terms of the punctuational model. This publication unleashed a vast, largely polemic, debate between those demonstrating gradualism (Chapt. 9) and those interpreting these examples by punctuation.

As S. J. Gould declared [8], it is quite evident that_ "The debate is not a struggle for life and death between gradual evolution and punctuated equilibria in which a case well substantiated in the one theory will invalidate the other. As always in natural history, we have to estimate the relative frequency of the various theoretically feasible possibilities. Gradual evolution, for example, dominates among mice, whereas the punctuational mode prevails among African wild pigs.

In his 1983 paper [8], S. J. Gould has softened his approach by admitting that, depending on the group concerned, one of these two models might apply in preference to the other.

We should remember that punctuation is a model and not a theory as some authors would like to make out. It is thus a "simplified representation" of the patterns and consequences of the model of speciation by geographical isolation, as proposed by E. Mayr, combined with a concept in which the species are considered as being stable.

Evidence for Punctuational Equilibria

The punctuational model was elaborated on the basis of two examples. The evolution of the trilobites [9] (appendix 11.1; Fig. 11.2), and of the Bermudas gastropod molluscs [10] (Figs. 11.3 and 11.4; Appendix 11.2) illstrate the points: first, the stability of the species in the palaeontological record; second, the fact that the most important morphological changes occured during the differentiation of specific new lineages; and third, the fact that new forms originated from peripheral isolates.

These examples support E. Mayr's model of speciation through geographical isolation with a reduction in the number of individuals. They also show that the differentiations took place quite rapidly and that the resulting forms are then globally stable, with their morphologies fluctuating under the influence of climatic variations. Here we also have the mechanisms for alteration of the developmental programmes which appear to be responsible for morphological transitions to other forms when environmental conditions became difficult for the organisms concerned.

These studies stress the important fact that the species succeed each other in the same area after having differentiated in the marginal regions. However, some small reservation has to be expressed to the extent that these examples only represent variations within a spatial and temporal species and do not represent the differentiation of truly new species by the acquisition of reproductive isolation.

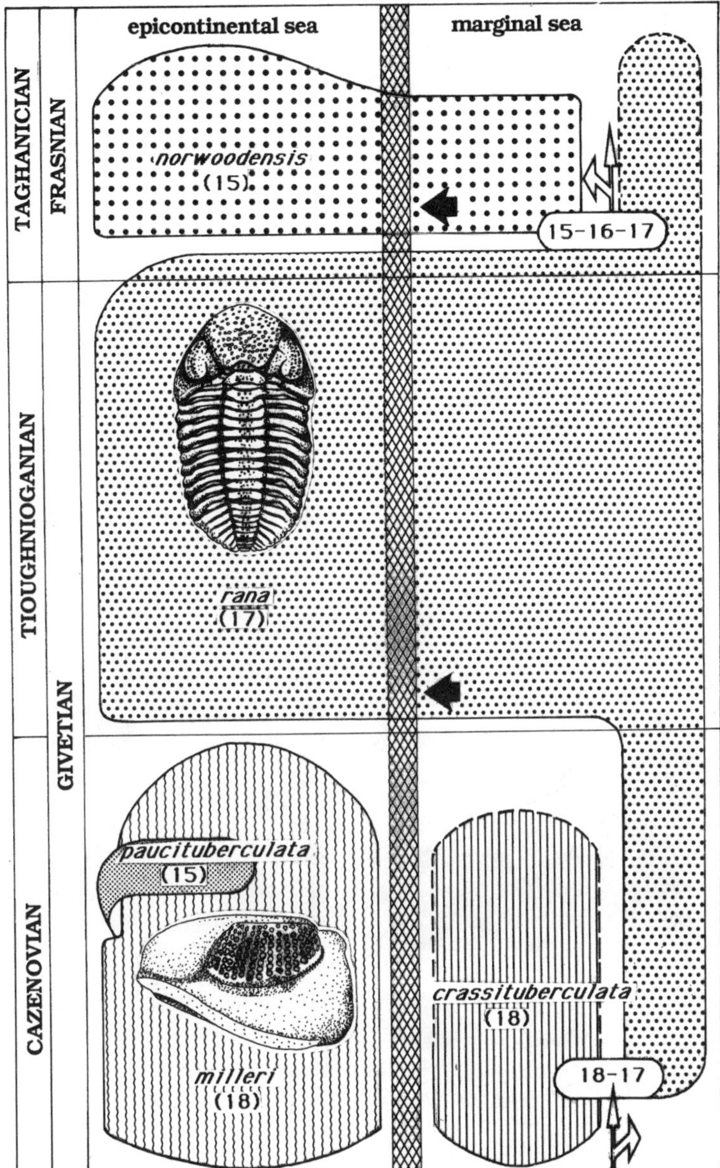

Fig.11.2. The Devonian trilobites of New York State. The succession of trilobites forms in the epicontinental sea cannot be explained by gradual evolution in situ, but is accounted for by the appearance of new forms in the eastern sea (marginal sea) during brief periods of speciation and their subsequent migration into the epicontinental sea during transgressive phases. Paedomorphosis appears to be the evolutionary mechanism operating in these trilobites. This example was used by Eldredge and Gould to illustrate their punctuational model. The trilobite eyes are made up of ocellae (like those of the present fly). *Numbers in brackets* indicate the number of rows of ocellae in the eyes, while the *white arrows* mark the formation of new morphologies through a reduction in the numer of rows of ocellae. (After Eldredge [9].)

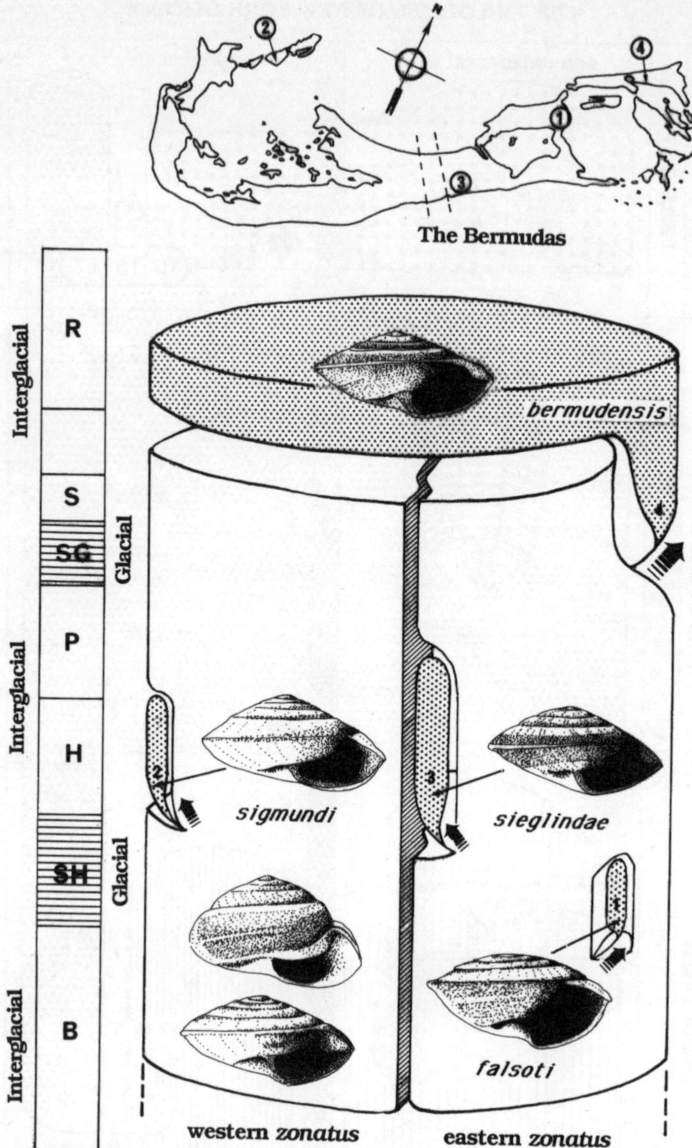

Fig.11.3. The Bermudas molluscs. This example, which has also been cited as evidence for the punctuational model, actually confirms the model of allopatric speciation by Mayr, with new forms appearing in peripheral isolates (1–4). *Below* Two forms of *zonatus*, one eastern and one western, from which local subspecies started to form in isolates 1–4. Their life span is variable. The forms are *falsoti (1)*, *sigmundi (2)*, *sieglindae (3)*, and *bermudensis (4)*. Here, paedomorphosis is the evolutionary mechanism permitting resident populations to survive under difficult conditions. *Above* Map of the Bermudas, with the numbers *1–4* marking the point of differentiation of the subspecies. A climatic chronology of the Bermudas during the Quaternary is shown to the *left*. Names of formations: *R* Recent; *S* Southampton; *SG* St. George; *P* Pembroke; *H* Harrington; *SH* Shore Hills; *B* Belmont. (After Gould [10].)

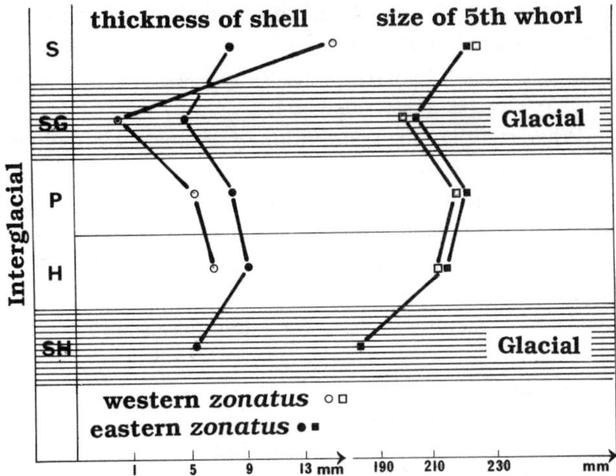

Fig.11.4. Climatic fluctuations vs. variations of certain features of the Bermudas molluscs. The morphological variations are small and it is impossible to distinguish a genetic component from an ecophenotypic component. However, it is possible to show a relationship between morphology and climate, as the variations in shell thickness and size correlate directly with climatic fluctuations. *S* Southampton; *SG* St. George; *P* Pembroke; *H* Harrington; (After Gould [10].)

Examples of stasis are tacitly admitted, although they are not always demonstrated by studies which fulfil the same rigorous criteria as those of gradual evolution.

Among living fossils, the present inarticulate brachiopod, *Lingula,* possesses a shell that is virtually identical to that of its Ordovician ancestor. Possibilities for variation of the shell of *Lingula* are limited to variations in width, which give to the shell a more spindle-like or ovoid shape. However, the internal structure of the soft parts of the ancestral form are not known and could perhaps have been different. This is an irritating problem of palaeontology, where the unknown relationship between information on the shell and on the entire organism will not allow us to prove absolute stasis.

One of the better documented examples for stasis is represented by the lineages of the archaic herbivorous mammals of the Tertiary Clark Fork basin of Wyoming [11]. Other examples are known from rodents where certain mice remained in stasis for 1.5–2 million years before becoming extinct [12].

Radiations and Extinctions

Palaeontology, shows that biological diversity has been subjected to extremely important fluctuations through geological time. Increases in diversity are referred to as radiations, and reductions are known as extinctions.

We know that the current biological diversity is difficult to determine, as is shown by the wide range of published figures. More than 2 million animal species have already been described, with possibly another million awaiting discovery, and Plant species number 1.5–2 million. In the past, biological diversity must have involved several hundreds of millions of species, now extinct, of which only about 300000 have been described so far.

When studying the distribution of organisms in the palaeontological record, one might think of it as a continuous regular expansion of the number of species resulting from their multiplication within the various body plans. However, this is not the case! If we evaluate the evolution of life on a larger scale, we note that the history of organisms is marked by the alternation of phases of radiation, during which species multiplied rapidly, with phases of massive extinctions which led to the disappearance of entire groups. The latter phases are referred to as bioloical crises.

The Major Crises of Life

The inventory of marine skeletal animals [13] shows that there were two important global phases of radiation separated by a period of numerical stability and a phase of extinctions at the end of the Palaeozoic. In addition to this, there were four more important crises (Fig. 11.5); thus, five massive extinctions are recorded by the palaeontology of marine organisms. To this should be added information from the terrestrial environment which includes the extinction of a large number of mammalian species during the Tertiary and after the last glaciation.

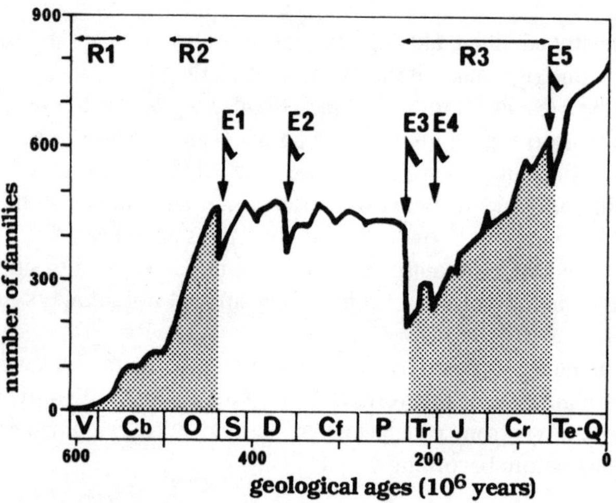

Fig.11.5. Radiations and extinctions through geological time. The variations in the number of families of marine skeletal organisms exhibit two large phases of diversification, one at the beginning of the Palaeozoic and the other during the Mesozoic. They are highlighted by three phases of radiations: *R1* upper Vendian-lower Cambrian; *R2* Ordovician; *R3* Triassic to upper Cretaceous. These radiation phases are terminated by periods of massive extinctions: *E1* end of Ordovician; *E2* end of Devonian; *E3* end of Permian; *E4* end of Triassic; *E5* end of Cretaceous. The extinction at the end of the Permian, when over 200 families became extinct, was the most important. *V* Vendian; *Cb* Cambrian; *O* Ordovician; *S* Silurian; *D* Devonian; *Cf* Carboniferous; *P* Permian; *Tr* Triassic; *J* Jurassic; *Cr* Cretaceous; *Te-Q* Tertiary – Quarternary. (After Erwin et al. [13].)

The first phase of diversifications started at the end of the Precambrian more than 600 million years ago and ended during the upper Ordovician (435 million years ago). It may be subdivided into two steps. The first, in which all the major body plans had been established by the end of the Cambrian (550 million years ago), led to the appearance of 120 new families of organisms. During the second step, up to the end of the Ordovician, another 350 new families appeared.

Diversification during the Cambrian was exponential, resulting from the mere multiplying effect of the continuous addition of new taxa. The rate of diversification of the individual taxon appears to have been constant, thereby excluding the influence of external factors, such as changes in the ocean chemistry, an increase in oxygen content, climatic changes or the appearance of predators, which have been invoked in various hypotheses to explain the situation.

If this growth had been maintained, then the seas would now be filled with more than 1 billion metazoan orders. However, as there are only 248 orders, we have to assume that certain limiting factors have operated. The decline in the rate of diversification, observed after the Ordovician, must have resulted from a concomitant increase in the rate of extinctions.

Analysis at the family level shows that there was a "Cambrian fauna" in pseudo-equilibrium during the mid-Cambrian, and this was replaced during the Ordovician by the expansion of the typical "Palaeozoic shell fauna" which compared with the Cambrian fauna, were made up of three times the number of families and four times the number of genera. The pronounced extinction, a decline of about 20% at the end of the Ordovician was followed by a state of equilibrium up to the end of the Palaeozoic.

Following the lower Cambrian diversification the environment appears to have become saturated, and the subsequent dynamic equilibrium results from the increasing rate of extinctions and the decrease of speciations in a saturated ecosystem.

This situation may be explained by a two-phase model with two independent but interactive groups of processes. The first phase corresponds to the Cambrian fauna entering a broad habitat with ample nourishment, whereas the second phase concerns the Palaeozoic fauna which appeared globally in a more specialized way, occupying rather narrow ecological niches.

From the start of the Silurian (435 million years ago) to the end of the Permian (235 million years ago), the diversity of organisms remained rather stable at about 500 families. However, this period was subdivided by two phases of extinctions, one at the end of the Ordovician (435 million years ago) and the other during the Devonian (350 million years ago). During these phases about 100 families disappeared, to become rapidly replaced by others in the subsequent phases of mini radiations during the Silurian (435–395 million years ago) and the Carboniferous (345–332 million years ago).

The end of the Palaeozoic was marked by the most important extinction of all, when more than 200 families of organisms abruptly perished. Some 13% of all animal classes disappeared, and in the invertebrates alone 17% of all orders, 52% of families, and 65% of genera disappeared. An evaluation of the number of species that became extinct is more difficult, but it appears that 88–96% of all species disappeared [14]. Thus, the invertebrates virtually vanished completely from the seas during the late Permian.

The second phase of radiations commenced in the Triassic (235 million years ago and has continued up to the present day, leading to the appearance of some 600 new families. However, this expansion was interrupted by two phases of extinctions, one at the end of the Triassic (195 million years ago) and one at the end of the Cretaceous (65 million years ago). During each of these phases about 100 families disappeared. The extinctions at the end of the Cretaceous, which included, in particular, the extinction of the dinosaurs, have greatly fanned public imagination. We shall return to this later.

Life on Earth shows a discontinuous history, where phases of extinctions alternate with phases of radiations. The phases of extinctions cleared numerous ecological niches which then became filled again in the subsequent phases of radiations. The pronounced diversification of families from the early Triassic onwards strongly suggests a multiplication of these ecological niches and increasing specialization of the various species.

One of the major problems inherent in these biological crises and extinctions is the question of their origin. They must be ascribed primarily to factors which are external to life itself and which could result from events in the history of the Earth, like the relative position of the continents, the creation of new oceans, or the extension or regression of shallow continental seas. Climatic changes could also have contributed to these crises. These phenomena constitute random events or accidents disturbing the advance of biological evolution.

Mathematical evaluations of the potential periodicity of these extinctions suggest a cyclicity with a duration of about 26 million years. This hypothesis, however, is much disputed.

The upper Cretaceous crisis, although not the most important, has always fascinated scientists. Among the numerous theories advanced, the cosmic catastrophe features prominently and is discussed below.

The Widespread Extinctions at the End of the Cretaceous: Collision With a Meteorite?

The end of the Cretaceous was marked by an important crisis and the disappearance of numerous groups of organisms, both in the seas and on the land. In the marine environment, molluscs like the ammonites, belemnites, inocerams, and rudists disappeared, as well as some sea urchins, bryozoans and brachipods. Among the marine reptiles, ichthyosaurs, plesiosaurs, etc. became extinct, as did a considerable amount of the plankton. On the continents, well-known extinctions included the dinosaurs and flying reptiles.

As a number of erroneous ideas have been proposed in the literature, it is necessary to clarify several points. The crisis at the end of the Cretaceous is by no means the only nor the most important one in the history of the Earth.

Detailed analysis shows that these extinctions are not abrupt, as is always too frequently assumed, but that they extend over a certain time span. The decline of the dinosaurs, for example, started several million years before the end of the Cretaceous, and the rate only increased during the final 300000 years. Some 60 theories have been advanced to explain this biological crisis. The great difficulty in finding a coherent explanation rests with the fact that while certain groups became extinct, others remained unaffected. This applies in particular to the fish, tortoises, crocodiles, lizards, snakes, birds, and mammals, all of which started to proliferate.

Among the theories advanced to explain the extinction of the dinosaurs are: they did not find enough food, or they devoured each other down to the last individual! An anatomical deformity, a defective physiology, or a hormonal disequilibrium led to their disappearance! Certain mammals developed a taste for dinosaur eggs! The large carnivorous dinosaurs exterminated the herbivorous ones, only to find themselves without further food! Some theories have invoked terrestrial catastrophes, such as the supposed formation of the moon out of what is now the Pacific Ocean, leading to earthquakes and tidal waves. An intensification of cosmic radiation, resulting from a supernova in the vicinity of our own solar system, could have led to lethal mutations. One can see that there has been no lack of imagination, but unfortunately science is unable to satisfy the requirements of "palaeofiction", although this type of literature sells rather well.

A new hypothesis involving a catastrophe was proposed by Alvarez and coworkers in 1980 [15]. They suggested that the Earth collided with a meteorite at the end of the Cretaceous. The theory is based on the discovery of enhanced levels of iridium in marine and continental sediments at the Cretaceous-Tertiary boundary at many sites throughout the world. The impact of this meteorite led to an opaque dust cloud that shrouded the earth in darkness for several months. The resulting lack of sunlight is thought to have caused the death of more "sensitive" plants and animals.

It should be borne in mind that iridium, although possibly of cosmic origin, could also be derived from deeper layers of the earth's crust. However, this explanation has been rejected by the followers of Alvarez' theory. Three arguments have been advanced in support of his theory; namely, the presence of microspherules resembling micro-meteorites, a layer of red clay attributed to the fall out from the dust cloud, and the presance of small "shocked" quartz crystals formed at the region of impact of the meteorite.

This hypothesis has led to numerous studies and, as a result, it has been increasingly contested in recent years. Hallam [14], in particular, has repudiated the supporting arguments and has proposed a number of counter arguments.

The palaeontological record shows clearly that the decline of the terrestrial vertebrates, and of the dinosaurs in particular, started well before the end of the Cretaceous and that only a few remaining groups disappeared abruptly at this point in time. A detailed study of the respective stratigraphic sections in Spain and Denmark shows that, for example, the ammonites disappeared well below the iridium-bearing horizon. As a consequence, iridium cannot have been the cause of their extinction. In the plant world, there appears to have been no catastrophic event causing widespread extinction at the end of the Cretaceous. However, there is an increase in the rate of disappearances in the Northern Hemisphere which may undoubtedly be ascribed to climatic cooling.

The indicative value of the supposedly meteoritic microspherules is somewhat doubtful as they could be of diverse origin. Some may have been derived from the filling of cavities in the sediments by algae during diagenesis and some could have originated from volcanic eruptions. Those rich in iridium could indeed represent true micrometeorites. The red clay horizon, which is not present in all profiles, could just as well have been derived from a concentration of non-calcareous material following a massive extinction of plankton. The "shocked" quartz grains could easily have been derived from volcanic eruptions similar to that of Mount St. Helens in the western United States in 1982. What is the significance of the iridium concentrations? In the section at Gubbio in Italy, in which this anomaly was first discovered, there are five successive iridium peaks, but no "shocked" quartz. This would require five successive meteoritic impacts over a time span of about 1 million years, and thus five impact craters.

Could the widespread marine regression at the end of the Cretaceous explain the disappearance of those marine invertebrate groups living along the continental margins? The approximately 200 m-deep continental seas extending from France to the Urals, for example, disappeared together with their faunas. The drop in sea level could have led to a more extreme and colder continental climate which may have led to the disappearance of the dinosaurs, as their eggs must have been rather sensitive to extreme temperatures. It could also explain the survival of birds and mammals which wére somewhat better adapted to such climatic changes by their ability to thermoregulate. Furthermore, the end of the Cretaceous was marked by a pronounced phase of volcanicity, culminating in the outflow of the thick flood basalts of the Deccan Trapp in India. This volcanic activity was responsible for the ejection of vast quantities of iridium- and sulphate-bearing vapours [16] into the lower stratosphere, which would, in turn, have resulted in immense quantities of acid rain with potentially devastating effects on plant life. The dispersal of volcanic ash could have speeded up the temperature decrease caused by the climatic changes. This type of volcanic activity is known to have started with a paroxysmal phase ejecting finely dispersed material derived from a great depth. The iridium content of this material could have been the initial cause for the disappearance of plankton and for the ecological disaster affecting terrestrial plant life, and the disappearance of plankton could have caused the extinction of those groups of animals that relied on it for nourishment.

This theory, although still not complete, is more plausible from a geological and palaeontological point of view than is the supposed impact of a meteorite. Similar processes could also provide an explanation for the other phases of marine extinctions observed during the Palaeozoic and Triassic.

Extinctions in Tertiary and Quaternary Continental Environments

The succession of mammals in North America appears to constitute a cyclic alternation of "diversity/residual diversity", a repetitive phenomenon with a periodicity of 2.3 million years [17]. The most important aspect of this is not the extinctions, but the formation of new communities from immigrants (climatic cyclicity?) and the ensuing repetitive evolutionary radiations. The sabre-toothed cats, for example, were replaced over the course of time by genera (*Haplophonus, Eusmilus, Barbourofelis, Meganteron,* and *Smilodon*), evidently occupying the same ecological niche. Rather similar cycles reconstructed from palaeobotanical data have been ascribed to climatic fluctuations.

The biological diversity of the continental environment has also suffered major fluctuations. Variations in Tertiary and Quaternary mammalian faunas have been analysed in detail, with special emphasis being placed on the great post glacial extinction.

This pronounced Quaternary extinction led to the disappearance throughout the world of about 200 mammalian genera which, for the most part, have not been replaced.

During the last glaciation, North America was covered by an immense ice sheet which isolated Alaska from the Great Plains. However Alaska was connected to Siberia by a land bridge of 1500 km that resulted from a lowering of the sea level by about 100 m.

South of the enormous mass of glaciers, in the actual United States, the distribution of organisms made up five animal and plant communities corresponding to the following faunal provinces: tundra, taiga, steppe-tundra, mountains in the midwest, and deciduous forests and savannah to the southeast. The groups of vertebrates then living in a certain territory were the same as today, although these groups were geographically separated [18]. This phenomenon is interpreted as resulting from a reduction of climatic seasonality during the last glaciation when the summer temperatures were reduced and the winter temperatures increased. This resulted in the exceptional formation of highly heterogeneous specific communities.

The increase in seasonal differences between 12000 and 8000 years B.P. destroyed these Quaternary glacial associations, leading to the establishment of the present faunas. The expanding deserts, prairie steppes, and hardwood forests became rapidly colonized by animals adapted to these habitats. This transformation caused the extinction of the large Quaternary ungulates including the horse, which was reintroduced only much later by the Spaniards.

The underlying cause of these extinctions was a climatic change, which led to a more rapid change of the vegetation than would have resulted merely from deglaciation. As a consequence, certain plant communities that were indispensable for the survival of these animals vanished. The changes entailed the loss of seasonal sources of nourishment, alteration of seasons favourable for reproduction, loss of predators, and loss of dwelling sites. Whereas the first palaeo-Indian hunters still encountered elephants, horses, and camels prior to 12000 years B.P., during the postclacial period there were only faunas impoverished in species but rich in individuals, such as the huge herds of bison.

On a worldwide scale, excessive hunting by man led to the disappearance of numerous species and genera, such as the large flightless birds living in islands, notably Madagascar *(Aepyornis)*, New Zealand (the moas), and Mauritius (the dodo).

The palaeontological and prehistoric data allow us to draw some valuable conclusions to assist our comprehension of the role of the environment in biological diversity. Pronounced fluctuations in biological diversity are anyhow characteristic of the history of life. Periods of extinction are caused by phenomena extraneous to the species themselves, in other words, by pronounced modifications of the environment.

Thus, after the postglacial phase of extinctions, we find a fauna which is impoverished in species compared with that at the end of the last glaciation. These extinctions were caused by climatic changes which led to the disappearance of plant communities indispensable for the survival of certain species. We can therefore say that the plant communities, which are controlled by the climate, are one of the main factors governing the success of animal communities. Furthermore, man represents an external factor causing extinctions by his hunting activities and his destructive influence on plant communities.

The periods of extinction are followed by periods of radiation during which the newly available niches become occupied. The last phase of extinctions should thus be followed by radiations of the same extent, unless man intervenes by modifying the various niches, which ist, unfortunately, the case at present.

Synthesis of Evolutionary Patterns: A Spatial and Temporal Model of Evolution

Our analyses have shown that there are several patterns of evolution. Gradual evolution has been well documented for a number of groups like rodents, early mammals, ammonites, etc. Morphological stasis is also present, especially among molluscs and trilobites. We should add to this the reversible and repetitive variations (or phenotypic plasticity) which, from a genetic point of view, correspond to a particular type of stasis. The respective species exhibit a wealth of morphological modifications resulting from a large potential variability. Of this, only a certain portion is expressed in response to the conditions prevailing in the environment at a particular point in time, varying in a reversible and repetitive way. This mechanism allows the species to adapt rapidly to environmental changes. It must be distinguished from true gradual evolution which results from cumulative variations according to a generally irreversible trend. This, however, also permits environmental adaptations. Figure 11.6 gives a schematic presentation of a change in shape with time, as implied by the three patterns observed.

Gradual evolution, combined with repetitive reversible variations, may be considered as a mechanism enabling a species that is not in equilibrium with its environment, to adapt itself. The phases of morphological stasis represent a certain equilibrium of the species with its environment, at least from a functional point of view.

As outlined in Chapter 9, stasis, gradual evolution, and repetitive and reversible variations may also be considered as late phases of speciation. Consequently, the evolutionary patterns actually correspond to the patterns of speciation.

Depending on the animal group concerned and on the history of the environment, these patterns may vary from one period to the next. The relative importance of these patterns has been shown for rodents in Eurasia [12, 19]. Gradual evolution appears to have been responsible for more than half of all morphological changes

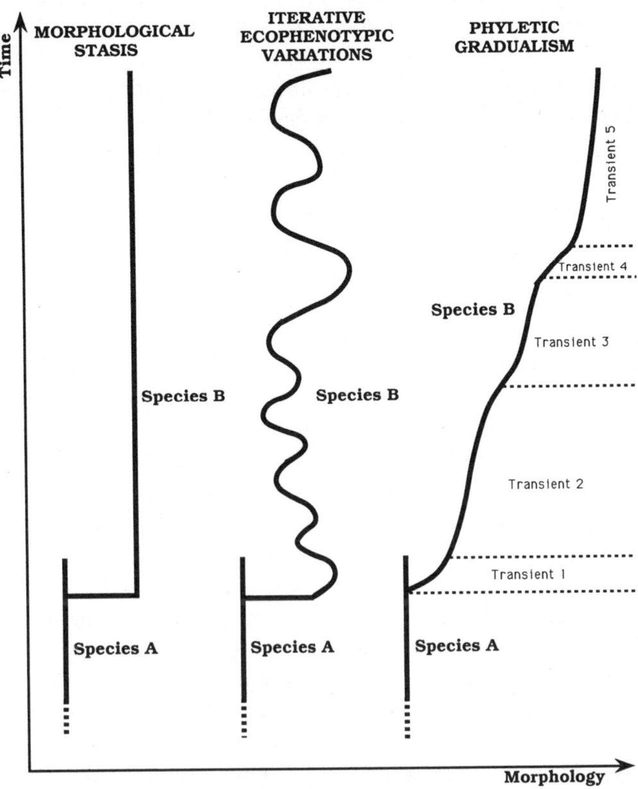

Fig.11.6. Modelling of evolutionary modes. Palaeontological data show that species originate rapidly on a geological scale, corresponding to a punctuation. Then species are either in equilibrium or disequilibrium. Morphological stasis occurs when internal constraints are stronger than external ones, although adaptations may develop at other organizational levels. In iterative ecophenotypic variations, the species is always in equilibrium with the environment because morphology oscillates gradually between two limits according to environmental parameters. Phyletic gradualism is the consequence of a disequilibrium between internal and external constraints, under directional selection. Irreversible morphological changes with time are described by palaeontologists as palaeospecies or morphospecies, and these are only successive morphotransients of a single diachronic species. (After Chaline and Brumet-Leconte [20].)

during evolution. On a relative basis within the lineages, gradual evolution accounts for 39% of recent evolutionary phenomena, the appearance of new species or lineages for 35%, stasis for a maximum of 25%, and reversible variations for 1%.

As has been aptly pointed out by S. J. Gould [8], we are not dealing with "a struggle for life or death between gradual evolution and the punctuated equilibrium in which a case well substantiated in the one theory will invalidate the other." Rather, the solution will be found in a pluralistic model integrating all the observtions [20].

Peripheral isolates containing reduced numbers of individuals appear to play a key role in the initiation of new species on a hierarchical basis which is more or less independent of the organizational level. Other evolutionary patterns among lineages, like gradual evolution, stasis, or repetitive reversible variations, or mechanisms such as hererochrony, together with innovations and multiple changes acting in successive phases, will permit the survival of newly formed lineages in a fluctuating environment over variable periods. Divergence leading to reproductive isolation facilitates the multiplication of the number of lineages and their coexistence. Just as a speciation represents a random event compared with the destiny of the respective lineage, so the execution of "possibilities" within the organizational plan by several successive species leads to an evolutionary trend within the lineage or to an adaptive radiation within a group. Even a poorly substantiated relationship between the multiplication of species and available ecological niches could assist with the diversification of groups.

Our model highlights the importance of the hierarchical concept of evolutionary levels and the spatial and temporal aspects of the formation of new species. The evolution of lineages appears to follow the data gleaned from modern biology and from the palaeontological record.

In answer to the question posed at the start of this chapter: evolution may be continuous and gradual within lineages, but discontinuous or punctuated during the origin of specific new lineages. The models of gradual evolution and punctuated equilibria explain some aspects of the observed phenomena and are a generalization of particular cases. Too hastily generalized, these models should give way to a more global approach, which represents an extension of punctuated equilibrium, but which also incorporates ecophenotypism and phyletic gradualism. Such a global model could be called "punctuated equilibrium/disequilibrium" [20], depending upon the relationship between the internal and external constraints which result in species living in equilibrium or non-equilibrium with their environment.

Chapter 12

Is Evolution Predictable?

This hotly debated question remains topical. Is biological evolution subjected to a strict determinism, to laws facilitating prediction, or is it at least in part subject to chance, and thereby unpredictable? We shall now consider this problem.

Are There Any Laws of Biological Evolution?

The concept of law, with its origins in antiquity, defined religious and civil obligations; however, the term was later extended to define certaine regularities observed within the Universe. Scientists study facts and try to elucidate the underlying causes or controlling factors. The concept of causality does not have to be defined in a law and this allows predictions to be made on the basis of the initial conditions. For a more in-depth analysis of these problems, the reader is referred to P.-P. Grassé [1], M. Delsol [2], and H. Tintant [3, 4].

When looking for the cause of a certain phenomenon, one usually produces a hypothesis by inductive reasoning, i.e. generalization from a restricted number of observations. At times, intuition will also play an important role in this step. The hypothesis will then be accepted or rejected on the basis of pertinent observations and experiments.

In science, there are two major types of law: mathematical ones, and those based on observations which frequently only have a statistical basis. This applies, for example, to the quantum theory of modern physics. Is it possible, however, to talk about laws in natural sciences? To answer this question, let us analyse and discuss a few instructive examples.

Linear Evolution or Orthogenesis

The existence of lineages exhibiting gradual evolution led certain authors to consider evolution as a phenomenon following a strictly linear, predetermined direction and progressing at a constant speed. This corresponds to “orthogenesis”, popularized by Eimer. According to the micropalaeontologist, Jan Hofker, [5] “orthogenesis always advances in a straight line. It is completely independent of sediment facies or deposit, and is expresses as time acting on the very patrimony of the lineage.”

This view represents a vitalistic concept of evolution in which any evolutionary phenomenon is controlled by an internal factor and is without any external constraints. It is a dogmatic concept of gradual evolution which may be expressed by the provocative statement of Pierre de Saint-Seine [6]: "Fossils encounter calculation".

This implies that evolution is a phenomenon controlled by rigorous determinism facilitating prediction. The palaeontological record shows effectively that gradual evolution does indeed exist, but that it "never" follows a strictly linear path and its rates change constantly as the result of fluctuations in the environment. Orthogenesis is a concept of the mind and therefore one cannot talk about the law of gradual evolution it seems to imply.

The Law of Increasing Complexity

This law was defined initially by C. Linné and then by J. B. Lamarck. The latter was of the opinion that after their appearance on Earth, organisms formed ordered sequences from simple to very complex forms. Some authors believe that Lamarck interpreted this increasing complexity as the expression of a pre-existing divine plan. This concept was revived by Teilhard de Chardin who asserted that the evolution of life proceeded to the development of a conscience and spirit. The palaeontological record actually tells us that although there is a global increase in complexity from invertebrates to vertebrates, this trend is by no means irreversible. The evolution of the leg of the horse, for example, shows that at some stage there was actually a trend for simplification. This also applies to those continental animals which colonized the subterranean environment or to animals which turned to a parasitic life style. Again, we cannot talk about a law.

Cope's Law of Non-Specialization Versus Depéret's Law of Specialization

The law formulated by Cope states that only non-specialized organisms are able to evolve further. Modified by Charles Depéret as the law of specialization [7], it states that once the major body plans had been established, evolution proceeded within these plans by trying out the various possibilities, eventually leading to forms with increasing degrees of specialization. The palaeontological record shows clearly that non-specialized types may diversify during phases of radiations, whereas extinction is the destiny of the more specialized forms. This trend, however, does not possess the status of a real law because, for example, as a result of heterochronies specialized forms may regain a less specialized status.

The Law of Size Increases, or Cope's Rule

This phenomenon has certainly contributed most to the elaboration of the concept of orthogenesis. It is based on the observation that, very frequently, groups or lineages starting with small forms exhibit a trend of size increases at highly variable rates, sometimes leading to gigantic forms. In most groups there are numerous examples of this trend; we have already mentioned the carnivores, horses, and titanotheres.

Although this observation was formalized by Cope, it was Depéret who gave it its general applicability to both invertebrates and vertebrates. The reasons for these size increases have been discussed at length, and range from the size of the respective ecological niche to a selective advantage. For example, for a vertebrate, which is able to control its body temperature (homeothermic), a larger size offers a certain thermodynamic advantage as a larger species requires proportionately less food (or energy) than a smaller one. Size increases should also be considered in terms of a correlation between the size and volume of an animal and its ecological niche. It should be borne in mind that when size increases by a factor x, the body surface area increases by x^2 and the volume by x^3. Newell [8] has compiled a number of examples for invertebrates.

Palaeontology provides proof of this phenomenon in a purely descriptive way, as we have shown in Chapter 9 with well-documented examples of gradual evolution and trends. However, is this sufficient for us to talk about a real law? Far from it, as in numerous groups or lineages we may observe exactly the opposite trend, namely, size reduction; this is illustrated by the dwarf forms in the equine lineage. Size is certainly one of the features that is most sensitive to environmental influence, and is therefore particularly susceptible to reversal; this is at least suggested by reversible repetitive variations.

Dollo's law of Irreversibility in Evolution

Louis Dollo, a Belgian palaeontologist, has shown that when an organ starts to evolve in a regressive sense, it will not be able to return to its original state [9]. This observation has become generalized as Dollo's law: Having started on a determined evolutionary trajectory, no animal may turn back. S. J. Gould [10] showed that L. Dollo realized, even in his days, that there was a problem of probability; in other words, unless a species, lineage, or group is re-subjected to all the previous events and unless all the responses to stimuli are exactly the same, it will statistically have "no chance" of returning to the ancestral state.

The palaeontological record, an immense field of data, shows that during evolution there is never an overal reversal of all the features of a lineage, but only sometimes a return of one or two earlier features [11]. One example of this is the reappearance of the second lower molar in the modern lynx, a feature that had disappeared from the group during the Miocene. Another example is a small tubercle in the upper incisors of the plesiapid monkey which, after its original appearance, has disappeared again from several lineages.

Can we then really talk about a law? Dollo's statement merely implies that the retro-evolution of a very complex assembly to an earlier state is statistically impossible.

A Law of Successions?

Palaeontology shows that large groups defined by a certain body plan begin with a weak presence which grows and leads to a peak before entering a phase of reduction. This eventually heralds the extinction of the group after a decline of variable duration. Based on this observation and on the intervals between the appearance of the major body plans, it appears that the main groups succeeded each other in time. Fish dominated the early Paleozoic, amphibians the Carboniferous, reptiles the Mesozoic, and mammals the Tertiary and the Quaternary. This is merely an effect of perspectives and does not correspond to a real superiority of one group over another. Fish, for example, were never more abundant than in the Tertiary when the radiation of teleost fish was much more pronounced than that of the mammals. Thus, we should not consider this perspective as a law, but only as an effect.

Does Chance Intervene in Evolution?

The role in evolution played by chance is a controversial one. Chance is incompatible with the strict deterministic system of the Laplace-type. The above considerations on the patterns of expression of various possibilities have shown that a strict mathematical law is not followed and that, as a consequence, exceptions are permitted. The element that disturbs evolution, by introducing contingency, which renders evolution unpredictable, is indeed chance.

What is Chance?
The concept of chance is rather complex and encompasses at least five definitions, ranging from subjective chance to absolute chance [2]. We will consider here the objective definition by Antoine Cournot [12]: Chance is the encounter of two causal sequence which are independent of each other.

This definition implies the pre-existence of independent causal sequences, the encounter of which leads to a random phenomenon that may be measured statistically. In complex determinism, we therefore refer to as chance those phenomena which cannot be predicted in detail.

How Does Chance Work?
Chance acts as a random stimulus initiating evlutionary processes, mechanisms, and patterns. It creates new starting conditions for the initiation of processes which themselves develop according to the respective circumstances until another random event intervenes to disturb the system [13].

One may say that chance leads to contingency, but that it is executed according to certain patterns. In biological evolution, contingency is expressed by the limited range of possibilities imposed by the structures of the living world. We could say that chance is only able to execute what is mechanically, technically, or functionally possible.

Extending this idea, we could say that natural selection results in a number of possible adaptations or in extinction. These are in accordance with, or are neutral in relation to, environmental factors which are also independent causal sequences interfering with the variability of genetic programmes. There is thus an interaction between external environmental factors and the internal constraints of the living organism. As P. L'Héritier remarked in 1954: "In total, selection has not created anything, it only chooses from the infinite possibilities the ramifying chain of creatures with have come into existence" [14].

The Levels of Chance in Evolution: The Lottery of Life

Chance acts at most levels of integration controlling the constitution of the animal. During the reproduction of an individual, chance may intervene in the encounter of sexual partners and during the formation of gametes, the first lottery of hereditary features, when the chromosomes are divided equally, forming the spermatozoids and egg cells. At this level it also intervenes during fertilization, the second lottery of hereditary features, when a single spermatozoid, out of the millions or billions produced, fertilized a single egg, leading to the combination of features defined in the genetic programme of the individual formed.

The genetic structure will be subject to random mutations, some of which are caused by external phenomena like ionizing radiations. These mutations result from molecular changes taking place in the DNA molecules. This may lead to errors which, however, take place within the deterministic framwork of biochemical reactions. As proposed by H. Carson, the genetic programme is made up of a "closed" region which is difficult to modify, i.e. it is less sensitive to mutations (these are the structural genes), and of an "open" part that is highly sensitive to mutations.

The variability in the structure of the genetic programme is actually expressed by an array of possible developments in which chance intervenes via the influence of environmental factors thereby controlling the appearance of the adult as shown by repetitive reversible changes. Because of this, development is not strictly and entirely controlled by the genetic programme (Chapt. 4). From all the programmes available chance will involve only those compatible with the environment. The genetic programme has accumulated an enormous quantity of neutral mutations over geological time. Under certain circumstances, some genes could activate the mutations which have so far been dormant, and this would lead to a new developmental programme that will be subject to the processes of natural selection.

When an individual develops, the random nature of its living conditions within a complex environment will decide whether or not it will reach sexual maturity and reproduce. At this stage, physical environmental factors, such as climate and geology, intervene and may lead to the disability or death of the individual. Other factors involved result from the biological environment of the individual within a population of its own species or from relationships with individuals from other species living in the same territory. We must stress the key role of predation, which may lead to the disappearance of many individuals within a population; these are usually the weak, less active ones, but also include those which through unfortunate circumstances are placed on the path of a predator. This is an important aspect of natural selection [15].

Chance and Necessity

Thus we see that chance intervenes throughout the entire life of an individual, population, or species by a sequence of events and accidents. Its action is rather variable and is felt by a species at various organizational levels through the initiation of processes affecting the genetics, biochemistry, physiology (functioning of organs), development, ecology (changes of ecological niches), or behaviour (ethological changes). Chance is also active at the community level on a macroevolutionary scale affecting genera, families, etc.

All these influences of chance, of external independent causal sequences exerted on the organism, disturb ongoing processes and mechanisms so that others, which may exist in the pool of possibilities contained in the genetic programme of the individual concerned, may be initiated. The various disturbances contribute to what we refer to as natural selection, and will introduce to evolution some type of contingency, rendering it unpredictable.

The strict position of the synthetic theory of the 1940s recognized neither the part played by chance with in the various hierarchical levels nor its significance for the patterns of expression of the possibilities of life. Furthermore, it did not take into account the fantastic possibilities opened up by heterochrony which, with a minimum of genetic modification, is able to lead to fundamental morphological changes and the establishment of an entirely new body plan (Chaps. 9 and 10).

Chance is therefore able to modify the initial conditions for the processes and mechanisms which are potentially present in the genome. However, the role of chance is limited by internal constraints, as the expression of possibilities is controlled by the structural and functional constraints to which the organism is subject. Necessity and chance are thus two complementary aspects of evolution.

Chance – A Necessity

Without the possibility of chance would life exist, if not as part of a universal plan, then at least as an inherent feature of matter? The very fact that life and the Universe possess a history in the form of a pathway which cannot be expressed by a mathematical law and is not subject to strict determinism, illustrates the unpredictable character resulting from the intervention of chance, random events, and accidents. What is the nature of these events? It is essentially the complex history of the oceans and continents, and the evolution of climates and environments which control this life.

Without chance, life may never have appeared on Earth. And also, without chance, life would not have experienced the tremendous evolution revealed by palaeontological studies. Chance thus appears to be one of the basic requirements for biological evolution.

Can We Talk About "Progress" in Evolution?

Until the times of Lamarck, natural scientists were content to describe the animal world as a linear scale of beings, progressing from imperfect (inferior) to more perfect (superior) states, and with man occupying the apex.

This nice, simple order started to become complicated and unsettled at the end of the 18th century when it was found to be necessary to admit to certain ramifications in order to describe the complexity of the living world. Fortunately, the real implications of fossils were not yet recognized. Even if there were branches, a single summit, occupied by man, could still exist. The presence of several summits would invalidate any idea of continuous linear progress. Would it be possible to compare, in terms of different degrees of progress, the top of one branch with that of another branch? This disruption of earlier resulted from the recognition of different, mostly independent body plans. The mollusc design is totally different from the insect design. Is one more perfect than the other? The question does not make sense; using more restricted designs, is a shark more perfect than a frog? Compared to the shark, does the frog represent evolutionary progress? All we may say is that it represents a change.

For the time being, let us abandon the concept of progress which is based on the judgement of values influenced by anthropomorphism. Let us talk instead about complexity: For example, an insect, fish, or mollusc is more complex than a sponge or jellyfish. In this case we are not dealing with a judgement, but with an objective statement of fact that is acceptable to all, and is independent of the interest we might have in one or other group.

Increases in the degree of complexity are, as we have mentioned, a global evolutionary trend, active since the appearance of the bacteria or even much earlier, from the very origin of life on our planet. This trend is revealed to us globally accross the panorama of the entire history of the organic world. When we descend to a lower level, to that of the actors in this history, the difficulties increase. A mammal is more complex than a mollusc or a cephalopod. But what is the position of the latter in relation to an insect? Although the two are completely different, there is nothing by which we can compare their respective complexities.

Sacculina, a crab parasite reduced to reproductive organs, is less complex than other cirriped crustaceans (rock and goose-necked barnacles) to which it is related. Anatomically it is ultrasimplified, but adaptively it is hyperspecialized. Indeed, it is hyperevolved in relation to other forms.

Thus, evolution is irregularly increasing complexity, but this does not necessarily represent progress. Should this latter term then be banned from the biological vocabulary? No, but it should be used with the utmost caution and under clearly defined conditions.

In the sense in which it is used here, progress represents a change in body plan which permits a more diverse and more complete utilization of all the environmental resources, resources signifying here habitats such as ecological niches. Increasing complexity can thus cause progress in evolution.

The sponges, as filter feeders, have become perfectly adapted since the Cambrian to their modest role and place, and have been able to maintain this position against competition from more elaborate filter feeders, such as lammellibranch molluscs (oysters, scallops, etc.). As such, the mollusc design represents an increase in complexity compared with the sponges. At the same time it represents a progress, as it gives this phylum the ability to occupy the filter feeder niche as well as other aquatic and terrestrial niches. The insects, on the other hand, have never really invaded the marine environment, confining themselves mainly to freshwaters and the continents which have offered a tremendous variety of niches. By their ability to utilize the resources of the planet, they also represent a certain evolutionary progress compared with the sponges. But in comparison with the molluscs? We cannot say, because we are dealing with two radically different types of construction.

The construction of mammals, and more especially their functioning, represents progress in comparison with the reptiles as mammalian reproduction (viviparity, placentation) and thermal regulation allows them to exploit an environment that is inaccessible to the reptiles, namely, polar climatic regions.

This leads to the biologically important conclusion that the term progress should not be included in a categorical statement such as "Evolution is progress". Progress may only be recognized step by step, at least between certain types of body plans. Its use therefore appears to be defensible only at higher levels of classification, like phyla or classes, but not at the level of genera or species.

Chapter 13

A Second Resumé: From Speciation to the Formation of the Major Body Plans

Biology tells us about the processes and mechanisms which are able to explain effectively the changes occurring in organisms and species. Palaeontology, with the temporal dimension at its disposal, adds to this data which are indispensable for our understanding of speciation and of the patterns and rhythms of evolutionary change. Only through this we are able to observe the historical development of the evolution of the living world as it took place through geological time. Based on data obtained from other branches of the earth sciences, palaeontology is able to evaluate the external constraints which frequently act at random on the living organism and which control the formation of new species, and their radiations or extinctions.

Redefining the Species Concept

From the point of view of many biologists, a species tends to be considered as a static entity of the present day. It is defined by systems of recognition between organisms to permit interbreeding between individuals of the same species, and to prevent crosses with individuals of other species by reproductive isolation.

One tends to easily forget that the recent biosphere with its 2 million living species actually represents a cross-section through an enormous cable made up of at least 2 million strands, each strand marking a specific lineage. Any evolutionist taking into account the essential of "time" will have to consider a species under the spatial dimension, that is its present distribution on the globe (the object of biology), and also under the time dimension.

This perspective must necessarily change our way of perceiving evolution. The species becomes a dynamic entity fluctuating in a "continuum" in space and time by the succession of generations. It thereby becomes a historical phenomenon. Because of this, the biological species concept should be extended into the time dimension, to become somewhat like "an entity continuous in time and space between groups of natural populations which, at any time during their history, are able to interbreed and which are reproductively isolated from any other specific analogous lineage."

This concept, which has already been more or less clearly stated by other palaeontologists, should integrate all the biological data discussed in the preceding chapters.

This approach will enable us to elaborate a new understanding of the formation of species, the basic phenomenon of evolution. We should, however, make a distinction between a concept of speciation and models of the actual formation of species.

Reconsidering the Concept of Speciation

The spatial and temporal aspects of the species, as defined above, entails important implications for the problems of speciation.

Initially, a species has a "starting point" when a group of individuals or populations acquires a new anatomy, either by the establishment of a new system of recognition between the individuals or by reproductive isolation from individuals of the parent species. It then has a certain "life span" corresponding to the extension in time of the genetic programme of the species. The new species has become a specific lineage that is more or less complex and which is subdivided to a varying degree into spatial subspecies at various successive points during the period of its existence. And eventually, a specific lineage comes to an "end" at the time of its extinction.

It follows from this that throughout the life of a species there is a continuity between successive generations within a population which assures the transmission of the genetic programme of the particular species. This genetic programme will accumulate mutations which either influence the organism during its development or, in the case of neutral mutations, are without effect. However, there will be no major discontinuities from the appearance of the species to its extinction when we stay within the same genetic programme.

As a result, a specific lineage has to be considered as a single species, despite the modifications it may have been subjected to during its history. This implies that morphologically defined palaeontological species correspond to sections at a particular instant through the continuity of a specific lineage, or to evolutionary stages within a species when it is modified to a varying degree over time.

Here we must admit – and it is only natural to do so if we are to take into account the temporal aspect of the species – that all the modifications which affect a species during its life span, and which result from the processes and mechanisms acting on the species from its origin to its extinction, may be considered as contributing to the ongoing formation of the species. This is in contrast to the usual biological concept of speciation which restricts the processes to the establishment of reproductive isolation.

A species may thus be considered as continually forming and transforming from its origin to its disappearance. Let us now look in detail at the main phases in this concept of speciation and at their characteristic points.

The first phase in the differentiation of a specific new lineage corresponds to the period leading to reproductive isolation from the parent species. Biological and palaeontological data suggest that peripheral isolates of populations represent favourable conditions for the formation of a new species. Why? The following explanations may be offered:

- Isolates exist under new ecological conditions around the periphery of a species and, as a result, may exhibit exceptional morphological variations.
- Due to the small number of individuals, they carry only part of the genetic programme of the species.
- In such small communities genetic drift becomes an important factor (Chap. 3).
- As a result of inbreeding, certain features may become fixed due to the duplicate presence (homozygote) of the respective controlling genes.
- Selective pressures are established which differ from those affecting the parent species.

Because of these factors, new species are more likely to appear in the isolates than in larger populations. However, their usual destiny is extinction, and those which result in established new species are rare. The first phase of speciation is a main object of biological research.

The second phase corresponds to the period following reproductive isolation which leads to the modifications that control the history of a specific lineage. Data for this, which accumulates over geological time, is supplied by palaeontological research. The palaeontological record shows that the future of a new lineage may develop in three main ways: by morphological stasis, repetitive reversible variations, or gradual evolution (phyletic gradualism).

Phases of morphological stasis may extend over different periods of time, sometimes covering several million years. However, the morphological stability of an organism, as understood by palaeontologists, does not imply stability at all organizational levels. Whatever the case, the existence of stasis bears witness to a state of equilibrium between the internal constraints of the organism and the external constraints exerted by the environment.

Reversible repetitive variations constitute a particular pattern of survival. Of all the potential variability in the genetic programme, only part will be executed as a function of the local environmental conditions. As these conditions change, so the morphology will also change, although it may be reversed if the environment returns to the initial conditions. We are dealing here with a very flexible process of adaptation to potential fluctuations in the environment.

Gradual evolution, the third mode, corresponds to irreversible changes in a specific lineage. The rate of these changes may vary over the history of the lineage, especially in response to fluctuations in the environment which act as stimuli. The transformations executed during gradual evolution lead to the formation of temporal species, recognized in palaeontology (phyletic speciation). This latter pattern was preferred by the synthetic theory, at least at its beginning, as it also recognized the existence of stasis.

When comparing the model of gradual evolution with the punctuational model, the problem of evaluating their relative importance in history becomes apparent and will vary according to the functional and structural constraints of the specific body plans, to the environments concerned, and to the actual circumstances. Among rodents and ammonites gradual evolution frequently occurs, whereas among molluscs, stasis and repetitive reversible variations are common. All the possibilities are developed by nature and any combination of the three extreme modes may exist. Evolutionary change may set in before or after reproductive isolation, and may combine in succession any of the three patterns.

When considering the temporal dimension of the species, these three modes determining the future of a specific lineage clearly suggest that modifications occurring after reproductive isolation cannot be distinguished from changes initiated within the parent species, and that they may thus be considered as late phases of speciation.

A species is therefore a dynamic, spatial and temporal entity in permanent formation and transformation.

The Models of Speciation

The classic models of speciation are of a geographical type, entailling the division of an ancestral population by a physical barrier into two or more populations in which the number of individuals will be reduced to a varying degree.

When an ancestral population is divided into two roughly equal parts that do not come into contact again thereafter, frequently a gradual divergence of the two groups will take place under the influence of regional factors. In this case reproductive isolation is of no signification and of no use. Although this dumbbell type does exist, it is rather rare in the palaeontological record.

When the ancestral population is divided into unequal portions we are dealing with a model of speciation by a temporal and spatial "bottle-neck", also known as the founder principle. The extreme example of this model would be a sole individual, a fertilized female, giving birth to a new species (consider the Hawaiian *Drosophila* species).

Are there also models of speciation that do not require geographical isolation? Extensive mutations ("hopeful monsters") are usually lethal for the individuals concerned and therefore would not be conducive, to say the least, to the formation of a new species. However, there is another possible, and more realistic, type of speciation, namely spontaneous reproductive isolation within a species as the result of a physiological or ethological mutation. This so-called sympatric speciation, although discounted by some authors, certainly plays a role among animals with poorly developed locomotory systems. White [1] has described this type of speciation in morabine grasshoppers of Australia as the stasipatric model.

A consideration of the temporal aspect of a species will reveal the existence in the palaeontological record of another type of speciation. In this case the isolate does not develop around the periphery of the area of distribution of the source species, but rather it develops through time as the species becomes confronted with an environment that is increasingly unfavourable. The species is then fragmented into residual isolates which, through the processes described above, will become susceptible to the modifications that initiate new phyletic lineages. Here again, reproductive isolation loses its significance because the parent species has disappeared.

Micro-, Macro-, and Megaevolution

The synthetic theory reduces evolution to a single mechanism (Chap. 2), responsible for variations of any extent. For some authors, this reductionist approach is not acceptable as they feel that the interaction between mutation and selection is only sufficient to account for the variations affecting the species within its own framework (subspeciation). This would be microevolution.

Diversification, the formation of new species, and thereby the establishment of the higher categories in systematics, like the genus, family, etc., would require a different form of evolution, that of macroevolution, for which the controlling mechanisms have not yet been elucidated.

Finally, Simpson distinguished a third pattern, that of megaevolution, which leads to the phyla, defined by the major types of body plan. This must have been active very early, mainly at the end of the Precambrian (570 million years ago) and was practically finished with by the end of the Cambrian with the appearance of the vertebrates, the last phylum.

Thereafter, evolution proceeded mostly by exploiting the potential inherent in these organizational plans. Diversification became manifest as evolutionary radiations, with the multiplication of species and, consequently, also of genera, families, etc.; these exploited to their utmost the potential of the environment.

These radiations represent the structural and functional responses of the organisms to transform themselves and also to utilize the many resources and ecological niches ("employment opportunities") available to them within the environment. Because of this, these radiations are referred to as adaptive. They follow each other through geological time. Among the more spectacular ones are those of the ammonites (Chap. 10) and reptiles during the Mesozoic, and of the mammals during the Tertiary.

We shall now outline the patterns and results of evolution. The question of the mechanisms responsible remains open. Is the reductionist position ("microevolution") defensible or must we rather adapt micro- and macroevolution, each with its own mechanisms of action on the genetic programme and each with its own directive force(s) enhancing selection at various levels? We do not yet know anything, about such macroevolutionary mechanisms. We must recall that the genome can engender any transformation from the simple "microevolutionary" change to the radical "macroevolutionary" remodelling that occurs in the four-winged *Drosophila*. At the genetic level there is thus no place for two different mechanisms. We agree with this view, which has also been defended by Dobzhanky et al. [2].

Determinism and Chance in Evolution

The title *Chance and necessity* was aptly chosen by J. Monod to stress the roles played by internal constraints, or necessity, and that part of indeterminism usually inherent in chance. Today, the title would be rather incomplete as the relationships between internal and external constraints are considered to be much more complex.

The internal constraints have been outlined in Chapter 6. According to an expression by F. Jacob, they entail a domain of possible and one of impossible realizations. Not everything is realized in nature! The constraints have the consequence, of channelling the possibilities included in the genetic programme. Studies of the regulatory mechanisms in genetics and development have shown that evolution may be continuous or discontinuous and that, despite the internal constraints, the potential expressions of the possibilities are so vast that no predictions can be made.

There are also external constraints interfering with the internal ones. For a long time it was generally believed that there were laws of evolution, like the law of increasing complexity, of non-specialization, or of succession. However, although biological phenomena, when taken individually, are subject to an inflexible determinism, as a whole they appear to obey statistical probabilities. Why?

This is because of the intervention of chance which constantly modifies the initial conditions. One should not, however confound determinism with a degree of predictability, as is all too frequently done. Chance! Much has been said about it; sometimes it is assigned the most important role and at other times it is completely rejected. We shall try to evaluate its role. Let us recall Cournot's definition of chance: "Chance is the encounter of two independent causal sequences."

Among the phenomena of complex determinism, subject to a multitude of factors as in the theory of evolution, chance signifies anything that is not predictable in detail.

Chance appears to act as a stimulus, initiating certain processes and mechanisms. Where does it act? The answer is clear: at the majority of organizational levels, as has been described.

It intervenes during the formation of the sex cells, constituting the first lottery of hereditary characters when the chromosomes divide to form spermatozoids and egg cells. It also appears at fertilization, when only one out of the million or billion spermatozoids formed is able to fertilize a single egg cell.

Chance intervenes again when the genome is subjected to mutations.

It also has a role in the influence of the external conditions of the local environment which control, to a varying degree, the ontogenetic or developmental pathways. It will thereby contribute to that part in the construction of an organism which is not immediately controlled by genetics (or epigenetics).

Chance is active throughout the life of an organism in terms of favourable or unfavourable environmental conditions, and the circumstances and incidents in life which decide whether or not an organism may reproduce. Environmental factors can be of a climatic, physical, or biological nature; the latter refers to the relationships between the organism and other individuals of the same or another species, for example, predation.

We see that continuously chance plays a role in the life of organisms and their evolution. It cannot, however, control everything. It acts within a set of limitations controlled by the internal constraints which channel the possibilities, and by natural selection. We have here the paradoxical situation that, on the one hand, chance interferes with environmental factors to contribute to natural selection while, on the other hand, natural selection, in partially controlling evolution, appears to act as kind of anti-chance, especially for mutations, and it is reinforced by the channelling effects of internal constraints.

Random events or chance disturb the prevailing processes and mechanisms by initiating others and by modifying, at each point of interference, the conditions of development or programmed evolution.

Chance actually introduces some contingency to evolution which, as a result, becomes unpredictable. One might say that without chance there would be no biological evolution and that, therefore, it appears to be necessary element for evolution, although by no means the only one. Chance and necessity, and the necessity of chance are the formulae which, with our present state of knowledge, best define their role in evolution.

Part IV
A New Approach to Human Evolution

Chapter 14

A Particular Type of Evolution – Human Evolution

"Man, This Unique Being"

This quote by J. Huxley reflects the present consensus between scientists, philosophers, and theologians alike, after a long period of fundamental disagreement on this crucial question for mankind.

The earliest writings on the place of man in the Universe date from the Sumerian and Akkadian peoples of Mesopotamia. Between about 3500 and 3000 B. C. the famous Gilgamesh epic was formulated, while the Atra Asis discovered on clay tablets, dates from the time of the Babylonian king Ammi Sadouqa (1646–1626 B. C.). A Babylonian poem on the creation, dating from the 11th century B. C. describes man as having been formed from a mixture of clay and the blood of a deposed king, from whom he also derived his soul. Man was created merely to take over the menials tasks of everyday life from the gods, and thus appears only as a slave or a toy to the whims of the gods.

The genesis of the Bible has been attributed to several authors, one of whom lived around the 10th century B. C. during the reign of King Solomon. Other parts were written by religious authors during the Babylonian captivity in the 6th century B. C. They formulated a new vision of man as having been created in the image of God, to rule over nature. This concept is much more optimistic and contrasts profoundly with the Babylonian view, especially in that it introduces an entirely new aspect, that of liberty [1].

Aristotle was the first to perceive man's originality, visualizing him as "a reasoning animal". Although an animal subjected to biological and physical constraints, man has acquired a unique position in the living world by the capacity for "reflective thought". Man's relation with divinity eventually became changed under the influence of astronomical discoveries.

"Renaturalized Man"

Up to the onset of scientific research at the end of the 18th century and beginning of the 19th century, theologians and philosophers perceived man as possessing superior intelligence. They considered man as a separate being, thereby maintaining the anthropocentric view of the Universe. In the western world, the origins of man were "explained" by a literal interpretation of the Bible, in which man is viewed as the result of a special divine creation.

Developments in palaeontology and biology, have forced scientists to reconsider man's place in nature. Man is now considered as the "most evolved primate" in the living world, perfectly integrated in the history of life, as reconstructed by palaeontology. This new scientific view of man was violently opposed by theologians during the 19th century, and especially after the publication of Darwin's book *The descent of man and sexual selection* in 1871, in which he showed that man is clearly related to the higher primates and derived from an ancestral form [2]. Opposition by the Church, and especially by Protestants and Anglicans who adhered closely to a literal interpretation of the Bible, was rather strong and still persists to the present day in the United States "creationism". In contrast to this, the position of the Catholic Church, was more pragmatic. In his encyclical *Providentissimus,* Pope Leo XVIII stated that there should be no conflict between science and faith and that concerning the emergence of man it is necessary to have recourse to science. This position was reconfirmed by Pius XII and John-Paul II: The Bible does not say how the sky was formed, but how we could reach it." We must make a distinction between the two levels of our knowledge, namely philosophy and religion, when searchin for an explanation of "why?"; this contrasts with science which tries to unravel the "how?", the way in which things come about. In this book we are dealing exclusively with ,,how", and we leave it to the readers own conviction to consider ,,why".

Some scientists have allowed themselves to be drawn to some type of ,,scientism" which claims to be able to explain by science. This erroneous approach is comparable with that of the creationists.

These extreme views, however, have by now been mostly abandoned. Man has been given his true place in nature, having been "renaturalized". However, he remains a unique being, through the acquisition of reflective throught, and it is this property which, as we shall see later, starts to modify evolution itself.

All recent biological and palaeontological findings have supported the interpretations given by C. Darwin.

Our Close Cousins in the Zoo

Man is a primate – this is quite obvious from zoological comparisons. On this basis, man was included next to the chimpanzee in the zoological classification by C. Linné in 1758. At the time, the only purpose of classification was to catalogue the various creations of the Bible; species were still considered as immutable. All the comparative studies of anatomy, molecular and chromosomal biology have effectively shown that the chimpanzee is the animal most closely related to man in modern nature. The gorilla and orang-utan respectively are further removed in this relationship.

The Biogeographical Message

The African chimpanzee is subdivided into two species, the common chimpanzee and the pygmy chimpanzee or "bonobo". The common chimpanzee occurs as three different, geographically separated subspecies: in Guinea (West Africa), on the right bank of the Congo River in Cameroon, Gabon, and the Congo (equatorial Africa), and in Zaire [3]. The much rarer pygmy chimpanzee is only found in Zaire, living in the forests on the left bank of the Congo River.

The gorilla is represented by two groups of populations which also live in geographically separated regions. To the west, from southern Cameroon to the Congo, we find the gorillas of the western plains or coastal regions. To the east, the gorillas of the eastern plains live in the mountain forests. There are also some mountain gorillas in the Virunga volcanic massif. These have been studied by Dianne Fossey who was recently murdered during fieldwork in the area.

The orang-utan is an exclusively Asian form. It lives in the Gunung Leuser National Park reserve in northern Sumatra and on the island of Kalimantan (formenly Borneo) where it is represented by two subspecies.

The geographical restriction of the chimpanzee and gorilla to Africa and of the orang-utan to Asia suggests a complex history.

The Molecular Message

Modern molecular biological research of the biochemical constitution of the primates has shown that the macromolecules of man and the chimpanzee are more than 99% similar, a truly amazing result.

In 1975, King and Wilson [4] compared the genetic programmes of the chimpanzee and man by analysing their amino acid sequences and immunological data. The two species are identical in their fibribopeptides, cytochrome C, and alpha, beta, and gamma haemoglobin chains. They differ, however, in other points and their small genetic distance corresponds to that observed between sibling species of *Drosophila,* or rodents. By hybridization of the nucleotide sequences, it can be shown that the proteins of the chimpanzee and man differ in 7.2 amino acid sites per 1000 substitutions. This effectively illustrates that the polypeptide sequences of these two species agree to more than 99%. This is confirmed by a comparison of the electrophoretic investigations of their proteins. In other molecular biological studies on the genes for globins psi to eta, an attempt was made to estimate the extent of the divergence between all higher primates [5]. This showed that man and the three large higher apes possess a high degree of identity; the observed divergence ranged from 1.61 to 3.52%.

The orang-utan is the most divergent of the higher apes, differing from the African apes and man by an average of 3.46% of the substitutions, over a range from 3.39 to 3.52%. The difference between the chimpanzee and man is the smallest (1.61%), while the chimpanzee and gorilla differ from man by 1.84% [5].

It is thus evident that the genomes of the higher primates are virtually identical. The small interspecific divergence between man and the chimpanzee amounts to about 2.5 times the variability within the human species (0.67%).

These studies show that the large apes (chimpanzee, gorilla, and orang-utan), which are grouped in the taxon Pongidae, do not represent a monophyletic group as the orang-utan is not directly related to man. Based on the molecular data [5], we may consider man and the large apes as constituting the unique family of hominids, which is divided into two subfamilies: the hominines including man, the chimpanzee, and gorilla; and the pongines with the orang-utan as the sole representative (Fig. 14.1).

All the biochemical similarities thus imply the existence of a common ancestor for man and the chimpanzee.

Based on the concept of a molecular clock, which assumes that the rate of mutation is constant over time, molecular biologists have made estimate of the respective times of divergence and on the rates of evolution within the various lineages. The time of divergence of the orang-utan branch from that of the large African apes and man is placed between 15 and 10 million years ago. The initial divergence between man and the chimpanzee on the one side and the gorilla on the other, took place between 8 and 5.3 million years ago. The separation between the lineage leading to man and that of the chimpanzee occurred between 7.1 and 4.7 million years ago. Recent data suggest that the rate of evolution in the human lineage was slower than that in the large apes; however these observations still await confirmation.

The Chromosomal Message: Proof of a Common Ancestor

Since the 1970s research on chromosome comparisons has achieved tremendous progress, and through new marking techniques it has become possible to unravel the sequential structure of chromosomes over 300–500 bands [6, 7, 8].

The present anthropoid apes possess 48 chromosomes whereas man has only 46. This reduction results from the joining together of two rod-shaped chromosomes in man to form the unique V-shaped chromosome referred to as chromosome 2.

A comparison of the number of chromosomal variations between the anthropoid apes and man shows that the chimpanzee is most closely related to man, as the two chromosomal formulae differ only in 10 positions. The gorilla differs from man and the chimpanzee at 14 chromosomal positions. The orang-utan differs from man in 14 additional positions, and from the gorilla and the chimpanzee in a further 16 and 18 positions respectively.

The fact that the gorilla, chimpanzee, and man probably have 11 non-modified and 7 modified chromosomes in common is further proof of the existence of a

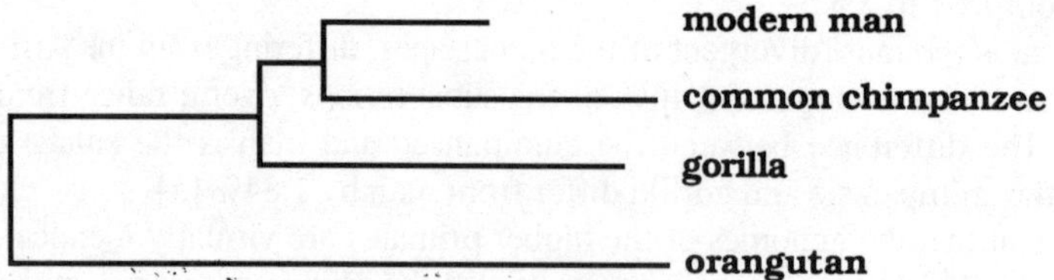

Fig. 14.1. Phylogenetic relationships between higher primates and man. The most parsimonious solution for phylogenetic relationships between modern man *(Homo sapiens)*, the common chimpanzee *(Pan troglodytes)*, the gorilla *(Gorilla gorilla)*, and the orang-utan *(Pongo pygmaeus)*. Branch lengths are in terms of the number of mutations. (After Miyamoto et al. [5].)

common ancestor, a homogeneous species not divided into subspecies. In addition, the possible existence of two chromosomes common to the chimpanzee and gorilla, and of three common to the chimpanzee and man [7] shows that the ancestral species must have split into three subspecies to account for these limited hybridizations (Fig. 14.2). One of the major conclusions drawn from this model is that

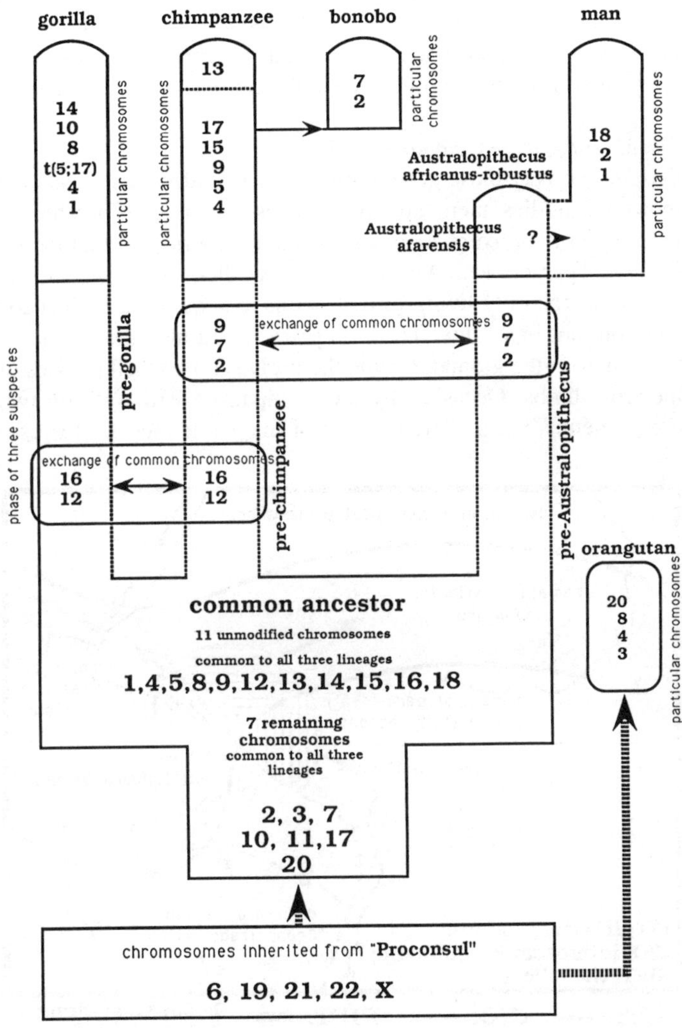

Fig.14.2. Chromosomal divergence between higher primates. Evolutionary scenario of pongid and hominid chromosomes. Variations in heterochromatin, which are common in numerous gorilla and chimpanzee chromosomes but are absent in man, have not been described in detail. Chromosome 20 shows 2 pericentric inversions with a common break point prohibiting evaluation of their chronology. (After Chaline et al. [9].)

one of these three subspecies must have been an archaic "pre-*Australopithecus*", i.e. a still non-bipedal ape. The transformation of these subspecies of the common ancestor into the present species is another domain for hypothesis. However, it is conceivable that climatic and environmental changes and geographical factors must have played a major role (Fig. 14.3) in the process [9].

The chromosomes, constituting a specific feature of the sequential arrangement of the genes, undoubtedly possess a certain variability about which still little is known due to the small number of available specimens. However, a comparison of the chromosomal formulae of species should enable us to delineate the parental relationships between the species, and to prove the existence of a common ancestor. For this purpose, we could reconstruct the chronology which must have marked the history of formation of the various chromosomal formulae.

The Message from Comparative Anatomy

Despite the very close genetic proximity revealed by molecular biology and chromosomal studies there are, nevertheless, such considerable morphological differences in the type of locomotion and skull structure that the situation appears to be almost paradoxical. An explanation is called for.

Man is the only living primate to have acquired a permanently upright stature and constant bipedality. The chimpanzee, and also the gorilla, remain on all four feet when on the ground, raising themselves up on the knuckles of the phalanges of the front limbs. Occasionally the chimpanzee will walk on two legs for several dozen metres, but the arrangement of the pelvic muscles forces it to return to nor-

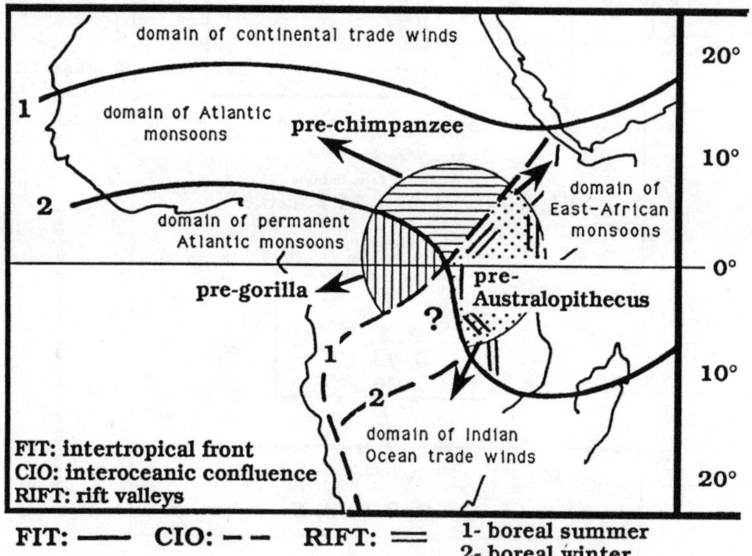

Fig.14.3. Schematic and hypothetical geographical repartition of the three component "subspecies" of the common polytypic ancestor. *FIT* Intertropical front; *CIO* interoceanic confluence; *1* summer; *2* winter. The equator was 7° more northely at 23 Ma, 5° at 10 Ma and 2° at 6 million years. (After Chaline et al. [9].)

mal quadrupedal life. The chimpanzee is also able to move by swinging with its long arms, a method described as brachiation. This profound difference in locomotion has important morphological consequences for the skeleton, affecting the shape of the pelvis and limbs, and the position in the skull of the occipital foramen (foramen magnum), through which the nerve cord connects with the brain and which controls the type of locomotion.

We have demonstrated that the two lineages are derived from a common ancestor. Their divergence is only very weakly expressed at the genic level, is somewhat stronger at the chromosomal level, and is stronger still at the morphological level. Evidently these morphological differences are directly linked to the different ecological niches, namely, forest and savannah, occupied by the two forms.

The Message from Development and Growth Provides a Clue

A comparison of the processes of growth and morphogenesis may show how the divergence between man and the higher apes came about. In 1926, Bolk [10] argued that human development may be distinguished from that of the chimpanzee by a process of foetalization. He considered that adult humans preserve features which are only present in the foetus or juvenile stage in the chimpanzee.

Until quite recently we were unable to understand how such divergences came into being. However, the discovery of the major role played by innovations and shifts in the developmental timing of morphological divergence (Chap. 9) will open another, so far neglected, avenue of research in palaeoanthropology. It should facilitate a connection between genetics and morphology through the intervening stages of developoment [11].

Recent research on the different types of craniofacial construction, which contributed to the development of hominization [12, 13], hase shown that the respective growth processes actually obery certain biodynamic laws. A. Dambricourt-Mallassé [12] has show that the history of the primates can be described by a succession of six models of development or "fundamental ontogenesis", relating respectively to the prosimians (tailed monkeys and tarsians), simians (flat-nosed monkeys of South America, and gibbons), large apes (orang-utan, gorilla, and chimpanzee), *Australopithecus,* archaic man *(Homo habilis, H. erectus,* and *H. neandertalensis),* and lastly modern man. The establishment of a new ontogenesis is effected by the structural readjustment following a divergence, i.e. subsequently to an undoubtedly complex assembly of coordinated changes in the timing of development.

Figure 14.4 illustrates the chronological shifts between the development of the large apes and that of modern man. It basically entails a slowing down of the various development phases in man compared with those of the large apes. The period of gestation lasts for about 8 months in the large apes but is extended to 9 months in man. This feature, which undoubtedly permits one or two additional multiplications of the neurons in the brain and thereby gives a larger cranial capacity, is eminently important to hominization.

A major point relates to the skull which develops through successive cranial and facial contractions. An increase in cranial flexure in the human embryo explains the disappearance of the curved canines, of the pronounced mastication sys-

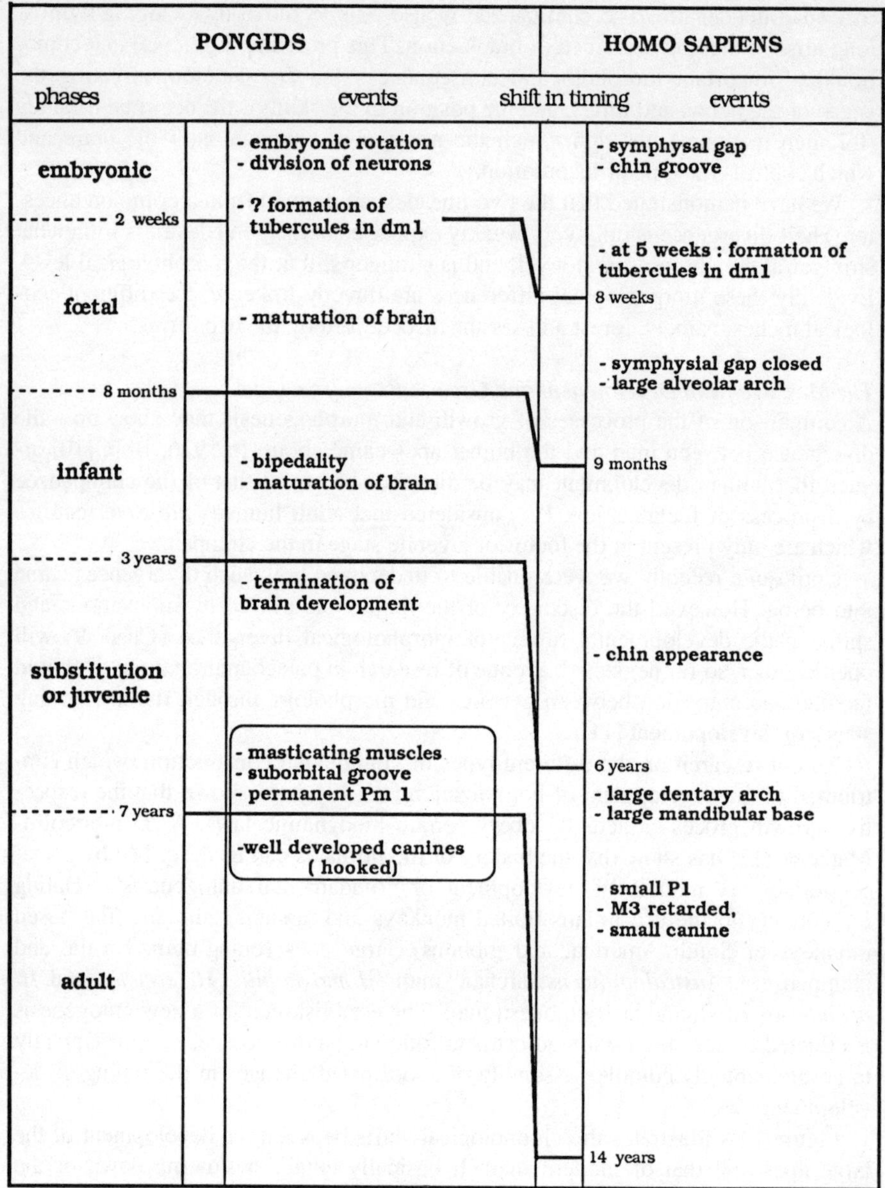

Fig.14.4. Chronological shifts between the development of the anthropoid apes and modern man. (After Deshayes and Dambricourt Malassé [13].)

tem in ancient man (jaws and associated cranial crests supporting the respective muscles), together with the appearance of a chin, and the disappearance of the brow ridges in modern man.

One of the most important consequences of hominization concerns the origin of bipedality. It is also present in the large apes just after birth but is lost with the appearance of the first molar after about 3 years. In the australopithecines and modern man it is maintained throughout life. Why? This remains a major problem, as yet unsolved.

These changes in the timing of development in man, compared with that of the large anthropoid apes, lead to a later onset of sexual maturation in man, who turns adult only after 14 years instead of after 7 years, as occurs in the higher apes. This facilitates a much longer period of learning for man with various cultural consequences, as will be described below.

In modern man, we thus observe a prolongation of the earlier phases and a regression of the later ones. It remains for biologists to look for the processes controlling these mechanisms of morphological change which explain the divergence between man and the large apes, and which do not involve extensive mutations. We now possess a clue for further explanation of the morphological evolution of the hominids.

The Message from Palaeontology, or a Parade of Common Ancestors

From the observed molecular divergences, geneticists have estimated that the common ancestor must have lived about 10–5 million years ago. However, the palaeontological record ist still rather poor, as strata of this age range are comparatively rare and have so far been little studied (Fig. 14.5).

The oldest hominids are the African *Proconsul* and the dryopithecines of Europe and Asia, existing 19–10 million years ago. Among these primates *Proconsul,* as the oldest example, may serve as a reference point. Its dentition is highly similar to that of the chimpanzee and the gorilla, whereas the cranial structure differs markedly from the latter two in the absence of the cranial crests.

Other apes, like *Kenyapithecus, Afropithecus,* and *Motopithecus* poorly known from a palaeontological point of view, may well represent intermediate stages on the way to the hominids. However, it would be premature to include them in a reconstruction at this stage.

There are no known fossil representatives of the chimpanzee or gorilla, except for a tooth found in 5-million-year-old sediments in Zaire. This has been assigned, with some reservation, to the gorillas. Because of the lack of further fossil examples, *Proconsul* and similar forms are now interpreted as a group of apes living in an environment similar to that inhabited by the chimpanzee and gorilla, namely, the tropical forests. As a result of similarities in feeding habit, they demonstrated morphological convergence with these two groups of apes.

Amongst the potential common ancestors of slightly younger age than *Proconsul,* the ramapithecines and sivapithecines occupied, until the 1980s, a prominent position as a result of studies by Elwyn Simons and David Pilbeam [14, 15]. The dental morphologies of these two groups exhibit numerous similarities with human dentition, including vertically positioned incisors, reduced canines, and the thick

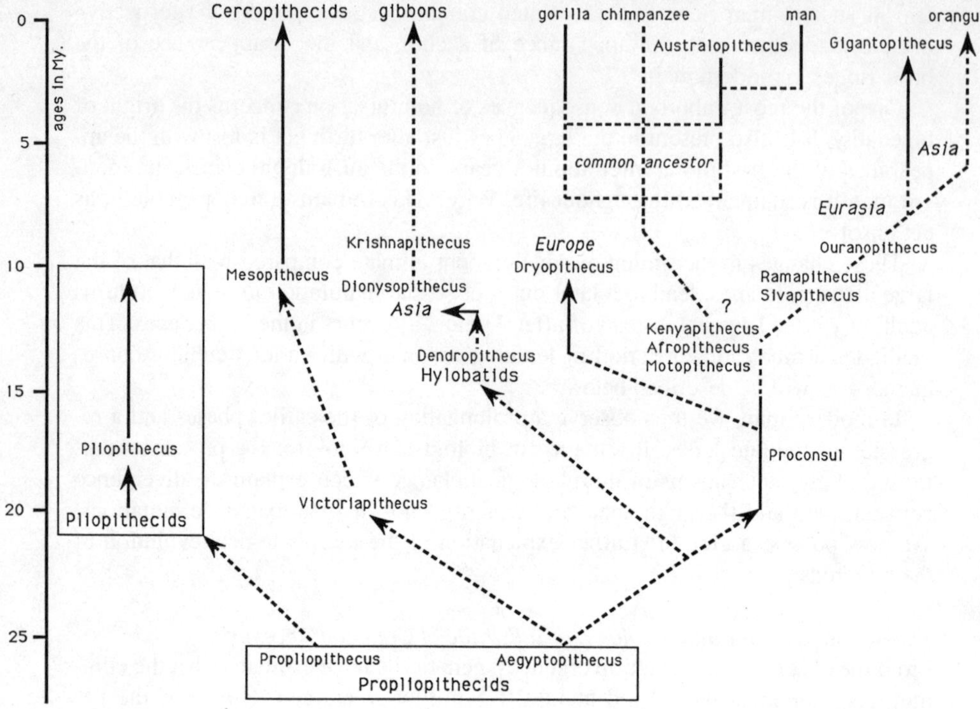

Fig.14.5. Phylogenetic relationships between higher primates according to palaeontological data. (After Mein [34].)

enamel cover of the molars. Differential wear on the molars indicates that tooth growth extended over a longer time span as a result of the lengthening of the juvenile period and the later onset of sexual maturity.

The ramapithecines and sivapithecines are present in 14- to 10-million-year-old deposits in Eurasia as jaw fragments and isolated teeth. Recently, a fragment of a sivapithecine face was discovered. The ramapithecines are smaller than the sivapitecines and it was realized that they were the males and females, respectively of the same species [16]. We now refer to them as ramapithecines. They are close to an ancestral form of the orang-utan [17]. The similarity of their dentition to that of man is now interpreted as the result of convergence.

They are also undoubtedly related to a group of giant Asian apes, the gigantopithecines, which became extinct during the Quaternary, and to *Ouranopithecus,* a primate discovered in Greece. The discovery in 1990 by Louis de Bonis et al. [18] of a cranial fragment with a massive concave face, like that of the orang-utan, tends to support its inclusion with the ancestors of the latter group [19]. The morphological similarities with man, as described above, appear to have resulted from convergent tendencies. Thus, the common ancestor must be sought elsewhere.

Oreopithecus, an ape famous during the 1960s, was discovered in a 7.6-million-year-old forest deposit in Tuscany Italy. Johannes Hürzelere proposed it as the potential ancestor because of similarities in the structures of the jaw and pelvis [20]. However, studies of the complete skeleton have suggested that we may, once again be dealing with morphological convergence, in this case of the jaw, teeth, and pelvis. This particular ape appears to have moved through the forest brachiation, as indicated by its very long arms. *Oreopithecus* has now been reassigned to an independent group of tail-bearing apes which includes the baboons and macaques. Recent studies may possibly reinstate *Oreopithecus* as a potential ancestor, but the relevant arguments are far from convincing.

Palaeontological data from 12- to 4-million-year-old rocks in Asia and AFrica are rare, partly because sediments of this time span are scarce and partly because political problems in countries like Ethiopia and Afghanistan, where such rocks do exist, render any fieldwork there rather hazardous.

Some tooth fragments have been discovered in deposits in the rift valley of Kenya [21] at Ngorora (12–4 million years old), Lukeino (6 million years old), and Lothagam (5.5 million years old). However, without a skull the teeth are difficult to interrpret, as was the case with the sivapithecines prior to the discovery of a facial fragment. The common ancestor, proven by molecular biology and chromosomes, has not yet been identified. The fossils from Ngorora, Lukeino, and Lothagam may well belong to the "prae-australopithecine" component of the common ancestor, as suggested by a comparison of the chromosomes [9]. If this hypothesis is true, then "prae-*australopithecus*" should possess a pongid or australopithecine cranial structure, but would not yet be bipedal. Future discoveries will be necessary to test this idea.

A common ancestor must undoubtedly be located in the 12- to 4-million-year-old sediments of Africa at the point where it gave birth to the chimpanzee and gorilla on the one hand, and to the australopithecines on the other hand; these forms are all of exclusively African origin.

The youngest known fossils, found in 4- to 1-million-year-old deposits, are the African australopithecines, bipedal apes which were already fairly far advanced on the way to hominization, and leading toward the archaic human lineage. Figure 14.5 illustrates the parental relationships between the higher apes and man as reconstructed from palaeontological and biological data.

What Do We Know About the Bipedal Apes?

The australopithecines, or bipedal apes, have been discovered in large numbers in African deposits of 4–1 million years in age (Fig. 14.6). They represent an important intermediate stage between the presumed common ancestor and the human lineage. They are generally ignored by some biologists, who tend to establish genealogical trees from recent data on biochemistry and chromosomes.

The australopithecine fossils are preserved in the ancient sediments of rivers and lake shores in East Africa, from Ethiopia to Tanzania along the African Rift Valley, and in caves in South Africa.

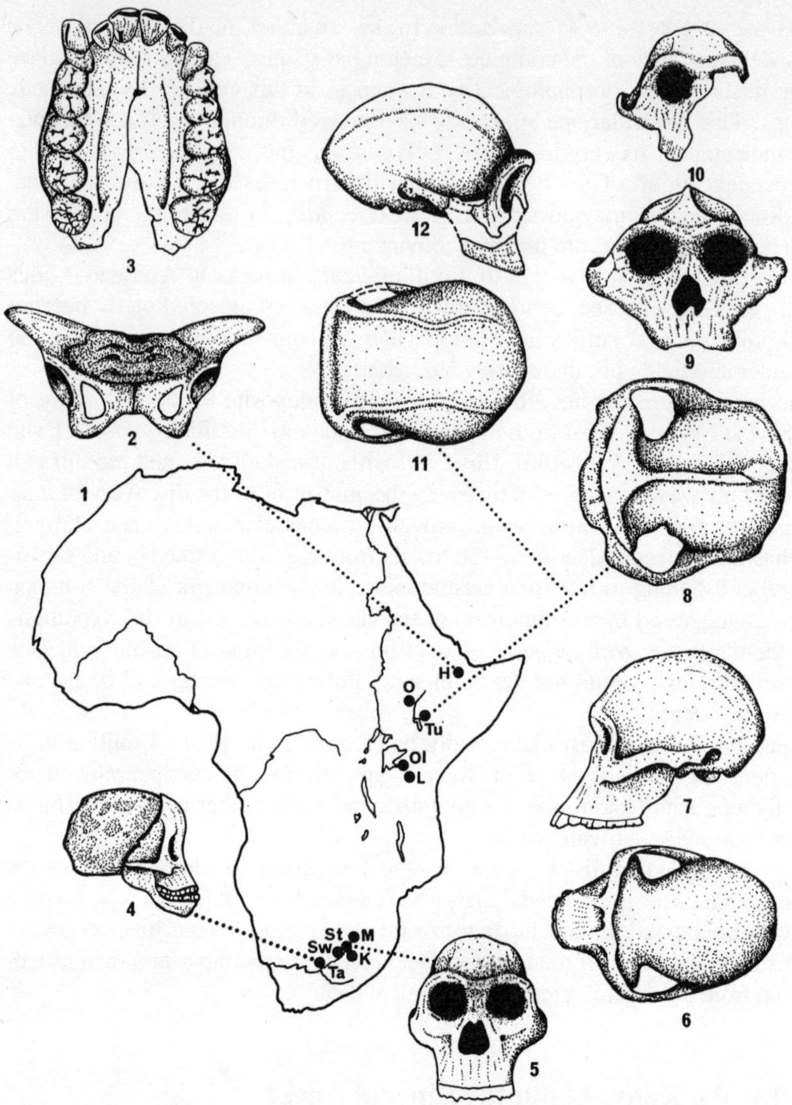

Fig.14.6. Australopithecines and man. *1* Map of Africa showing the main sites for australopithecines and primitive man (*H* Hadar/Ethiopia; *K* Kromdraai/South Africa; *L* Laetolil/Tanzania; *M* Makapansgat/South Africa; *O* Omo/Ethiopia; *Ol Olduvai/Kenya; St* Sterkfonte in South Africa; *Sw* Swartkrans(South Africa; *Ta* Taung/South Africa; *Tu* east side of Lake Turkana/Kenya).*2* and *3 Australopithecus afarensis: 2* Lucy, reconstruction of pelvis; *3* upper jaw – A1.200.1 – illustrating the reduction of the second lateral incisor, the rest of the jaw being similar to that of the chimpanzee.*4–7 Australopithecus africanus: 4* Taung child; *5–7* front and lateral views of skull of Sterkfontein adult ST5, undoubtedly a female.*8–9 Australopithecus robustus:* Et.406, undoubtedly a male in top view (*8*) and front view (9); note the pronounced sagittal crest and hour ridge characteristic of the male anthropoid apes.*10 Australopithecus robustus:* ET.732, a female form resembling ST5 from South Africa.*11* and *12 Homo habilis:* ET.3733 in top view (*11*) and lateral view (*12*); note the disappearance of the sagittal crest and the increase in cranial capacity.(*2* After Berge et al. [35]; *4–6* after Howell [36]; *8–12* after Walker et al [37].)

The most popular of these fossils, “Lucy“, discovered in the Afars of Ethiopia, consists of a pelvic region with a short iliac bone, like that found in man. Dated as about 3 million years old, it proves that the oldest australopithecines had already acquired an upright posture and a bipedal gait, an important innovation maintained in the human lineage. This feature may thus be considered as a characteristic of the first phase of hominization.

This interpretation is solidly underlined by the discovery of a track of hominid footprints, fossilized in the ash of a volcano that erupted about 3.6 million years ago. These oldest australopithecines are now referred to as *Australopithecus afarensis,* while the younger 3- to 2-million-year-old specimens are known as *Australopithecus africanus.* The youngest representatives, 2–1 million years old, are grouped together as *Australopithecus robustus.*

The pelvic construction of the australopithecines shows them to be the first hominids by virtue of their having acquired an upright posture and bipedality. However, their skull still exhibits a number of crests and, on the basis of their dentition, they still resemble the chimpanzee and gorilla. Because of this, they could justifiably be called bipedal apes. The situation could result from the chronological shift in development of a “higher ape” type leading to the australopithecine model [11, 12, 13]. However, the australopithecines preserved their long arms, as shown by the remains of Lucy, and by a cubitus discovered by Y. Coppens in Ethiopia.

The interpretation of the australopithecine fossils has been marked by a polemic because of the implications of these remains for the very origin of man. Initially, the remains were described under a purely typological species concept, each fragment being assigned a generic and specific status. As a result, this is now of little systematic value. Thereafter, the fossils were considered as belonging to several species, but no account was taken of the great morphological variability of the anthropoid apes. The pronounced sexual dimorphism of the latter was also not considered. Thus, the features used to distinguish between the species were the very same as those which characterize the males and females of present species. Instead of interpreting the australopithecines as biologically abnormal, we should rather, based on our present state of knowledge, consider them as constituting a single lineage in which the three above-mentioned australopithecine species represent three successive evolutionary stages containing both males and females [11]. However, the informed reader will note that this concept is not supported by all specialists in the field and that in the literature quite different interpretations have been presented; from a biological point of view these may be more or less credible [22].

A Scenario for the Origin of the Australopithecines

To understand the appearance of the particular pelvic structure of the australopithecines that facilitated bipedality, we most compare the development of the chimpanzee and man. Recent data show that the shortening of the iliac bone is a precocious phenomenon during the development of the human foetus, taking place at least before the third month. It appears to be a consequence of the restructuring of the skull [11, 12, 13]. Furthermore, bipedality appears to be present in the early stages of development in the higher apes (Fig. 14.4); however they lose this potential at the time when the first molar appears, at an age of about 3 years [13]. As the iliac bone serves as a support for certain muscles, it is not inconceivable that the development of the pelvic and thigh muscles contributed to, and shomehow controlled, this modification of the morphology of the iliac bone. This remodelling must have appeared suddenly as one cannot envisage any intermediate morphological stage that, being non-functional, would have been tolerated by natural selection. This innovation represents an abrupt modification of the developmental programme of the iliac bone.

It could be imagined that this innovation appeared within a group of 10–20 quadrupedal individuals of the "pre-*Australopithecus*" subspecies, one of the presumed components of the common ancestry suggested by a comparison of chromosomes [9]. The possibility of remaining permanently upright gave the bearers of this innovation an immediate advantage over fellows group members, with such individuals thereby achieving the status of a dominant male or female.

In a society of dominant males and females, this phenomenon would have become established by an increased abundance of descendants and by the rapid inclusion of this feature in the genetic programme of the group with a pronounced tendency for inbreeding.

Once the innovation had appeared, the australopithecines started to evolve like other species, that is, by a slight increase in size. This could be interpreted as a tendency to reinforce adult features by the delayed onset of sexual maturity, thereby extending the growth period [11, 12, 13]. The australopithecines disappeared about 1 million years ago; however, before this, over a period of about 2 million years, they gave birth to the human lineage in the true sense.

What Do We Known About the History of Man?

Homo habilis, the oldest representative of the human lineage, dates back to 1.8–1.6 million years ago (Fig. 14.6). Its name (genus *Homo,* species *habilis*) signifies an ability to produce tools. It is characterized by a brain volume of about 800 cm^3, compared with a range of 400–750 cm^3 for the australopithecines. From well-preserved skulls we know that the occipital hole, through which the brain is connected to the nerve cord, is in a more anterior position than in the australopithecines. This indicates that *Homo habilis* moved in a more upright position. Furthermore, bony ridges, like the sagittal ridge that characterized the australopithecines, had disappeared. Man exhibits a trend of reduction of the jaw and dentition. We are dealing here with the separation of the development of "archaic man" from the ancestral development of the australopithecines [12, 13]. The Olduvai skeleton, OH 62, discovered by Donald Johanson and Tim D. White, suggests that the oldest specimens of *Homo habilis* preserved the long arms of their australopithecine ancestors, provided that the fossil has been correctly identified.

The next stage in this lineage is presented by erect forms *(Homo erectus)* which spread to Europe and Asia. The brain volume continued to increase eventually reaching 1300 cm^3. The pelvis of this erect man is morphologically identical to that of modern man. The shape of the iliac bone does not differ from that of the australopithecines, but it became rotated, thereby enlarging the pelvic cavity. This new construction facilitates, during birth, the passage of a larger neonate than that of the australopithecines, and it also allows a more efficient bipedal gait.

The lineage of erect man then underwent a divergent evolution in Europe and Asia, resulting in two geographically distinct contemporary subspecies. In Europe, the specialized group of Neandertal man *(Homo sapiens neandertalensis)* gradually developed, with a brain volume reaching 1600 cm^3. At the same time in Asia, the Middle East, or East Africa, modern man gradually evolved *(Homo sapiens sapiens)*. Around 30000 years ago, modern man started to arrive in Europe, replacing the Neanterthals which rapidly became extinct over about 1000 years.

The human lineage thus appears to be a unique lineage made up of the three species "*habilis*", "*erectus*", and "*neandertalensis*", occuring as three successive evolutionary stages. Studies of development [12, 13] suggest that the model of "archaic man" underwent a new phase of divergence to give rise to modern man. This resulted in an increase in the rotating and flexing movements of the skull, the development of a chin due to an increase in alveolo-dentary growth as compared with that of the rest of the mandibule, and in the disappearance of the brow ridges.

The main characteristics of the evolution of the human lineage are: improvement of the upright posture and of bipedality from *Homo habilis* onwards, through rearrangement of the iliac bone; and a gradual continuous increase in brain volume. Comparing the phases of development in the skulls of the chimpanzee, gorilla, and modern man shows that the skull shape of modern man results from a decrease in their rate of growth, referred to as neoteny, or foetalization by Bolk [10]. Although this slowing down could explain the round shape of the human skull, it does not explain its larger size. The delayed onset of sexual maturity facilitates a longer growth period (hypermorphosis) and, amongst other consequences, a lengthening of the learning period [11, 12, 13]. The preservation of the juvenile skull features explains not only the shape, but also the fact that the nerve cells can continue to divide for a longer period, resulting in increased cell numbers. This explains the greater volume of the human brain and its new mental capabilities. Such evolution implies dissociated heterochronies, with some traits being paedomorphic, and others peramorphic [23].

Man also differs from the chimpanzee by having acquired articulate language. Recent investigations by Jeffrey T. Laitman and Edmund S. Crelin [24] have shown that man was able to develop articulate language because the larynx is situated farther back than in the chimpanzee. This anatomical situation facilitates a size increase in the pharynx, enabling man to emit modulated sounds. the chimpanzee is unable to produce such sounds because, due to the small size of the Parynx, it is mainly restricted to using the lips and mouth. In modern man the larynx of the foetus and child, up to the age of about 1 year, is situated in a fairly high position, like in the chimpanzee. Only with the migration of the larynx to the base of the neck during the second year is the child able to modulate sounds properly. Once again we are dealing with a complex modification of the developmental programme of the higher apes.

What Are the Implications of Human History on the Theory of Evolution?

The palaeontological history of the hominids is of an exemplary nature as it enables us to demonstrate a number of observations of general importance to the theory of evolution.

The first point concerns the abrupt appearance of lineages in the palaeontological record and by which the species species represent discontinuities with their ancestral species, at least in the case of man. This clearly demonstrates that the palaeontological record mostly reveals well-differentiated species with many individuals, but that it "overlooks" those populations with a reduced number of individuals in which the processes of speciation are taking place. In these populations, mutants bearing innovations that have accumulated in the genetic programme over a considerable time may become established because of the small number of individuals. They correspond somewhat to the "hopeful monsters" of R. Goldschmidt, without having the extent of a systemic mutation as Goldschmidt visualized. Because of the small number of individuals, such innovating populations do not possess any statistical chance of becoming fossilized. In this sense, the appearance of new lineages corresponds, on a geological time scale, to "punctuations", irrespective of the rapid processes involved, and impossible to recognize.

Once differentiated, the lineages evolved "gradually" to become australopithecines and humans.

Hominid evolution thus suggests a model in which "punctuations" were followed by pronounced phases of gradual evolution (Fig. 14.7.). The morphological changes that took place during the phases of gradual evolution were as important as those that occured during the punctuations. This implies a modification of the punctuational model of S. J. Gould and N. Eldredge by introducing the pattern of gradual evolution to a model of equilibrium/disequilibrium [25].

The origins of lineages and their eventual gradual evolution appear to be the results of abrupt changes or of gradual modifications in the developmental pathways of ancestral forms, the so-called fundamental [12]. Embryological innovations and changes in the timing of development of ancestral forms play a major role in the establishment of lineages and in the morphological modifications characterizing their history. For the australopithecines, the profound innovations is represented by the shortening of the iliac bone and by a rearrangement of the supporting muscular system. Undoubtedly, this is the consequence of a divergence of the developmental pathway of the australopithecines from that of the higher apes; this innovation also led to an increase in the occipital flexion of the skull and to a forward displacement of the occipital foramen. It manifests itself further in a gradual increase in size and by a reinforcement of the adult features. For archaic man, this innovation corresponds to a reduction of the osteo-masticatory system and in the start of a gradual increase in brain volume, together with the preservation of a juvenile morphology of the ancestral form. It is accompanied by other shifts in development, such as those affecting the larynx which, in turn, permit an articulate language.

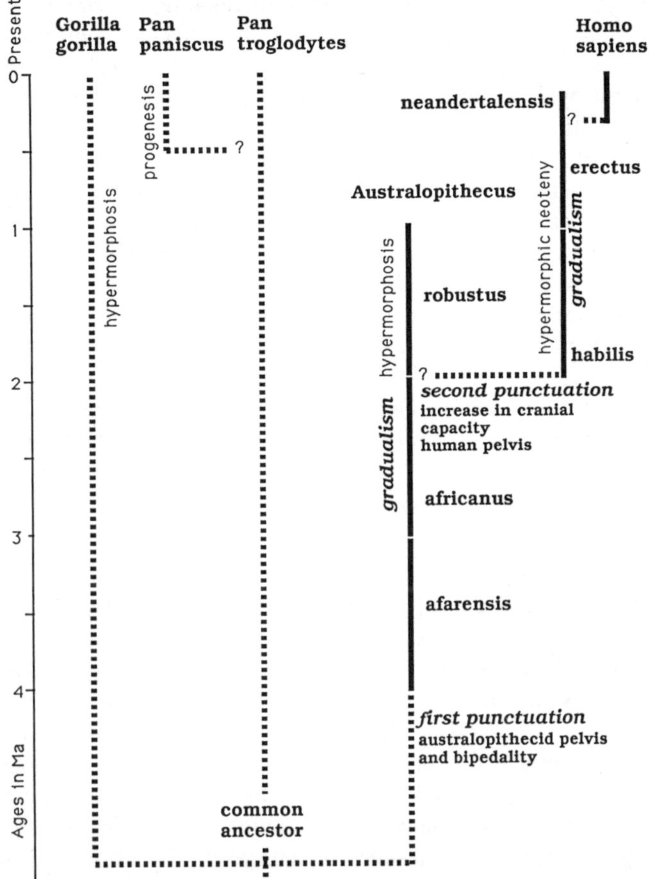

Fig.14.7. Scheme for the new theory of evolution of man and the anthropoid apes. It illustrates the differences in the structure of the pelvis, the role of changes in development, and underlines the parts played by gradual and punctuational evolution. (After Chaline et al. [11].)

The major role played by heterochronies in development offers a simple explanation of the apparent paradox of the genetic proximity between the chimpanzee and man. We can now understand that a very slight genetic modification in the developmental programme may result in wide-ranging morphological effects.

The differentiation between the anthropoid apes, and especially between the common chimpanzee, its pygmy form, and the gorilla, may also be explained in terms of such deviations in development (Fig. 14.4). Compared with the common chimpanzee, the pygmy form is caused by the precocious onset of sexual maturity which inhibits further growth (progenesis). The gorilla, in turn, results from a delay in the onset of sexual maturity which postpones the end of the growth period (hypermorphosis) [10, 26].

This example clearly illustrates that there is a hierarchy in the organization of animals from the genetic programme through development to the adult stage. This tendency is explained by the process of "canalization". It supports the concept proposed by F. Jacob, who stated that "the possibilities are realized by exploiting a virtual tinkering of previous structures."

This "tinkering" results partly from changes in the timing of development and from their extensions in time when certain constraints control the number of "potential morphologies". In conclusion it may be stated that the appearance of lineages seems to be unpredictable, relying on random circumstances, such as the appearance of a developmental anomaly. When it is immediaely advantageous to its bearer, as in the case of bipedality in the australopithecines, it will be retained by selection, but, in turn, it will be eliminated when it is disadvantageous. This, in broad terms, represents the theory of human evolution presently being developed.

The origin and development of the human lineage thus takes place under the same biological processes which control other vertebrate species. However, there are certain features – notably the acquisition of reflective thought – which set apart the human lineage in the history of the living world.

What Are the Stages of Development of the Human Psyche?

A species masters its own evolution! As aptly put by E. Mayr [27]: "One should not make the most tragic mistake of considering man purely as an animal. Man is unique; he differs from all animals in a number of features such as language, traditions, culture, and a very long period of growth and of weaning from his parents".

While language and the length of the period of learning are linked to biological processes, the development of a culture undoubtedly represents an innovation setting man apart from all other animal species. This novelty is a direct consequence of the increase in brain volume, in combination with the extension of the period of development which permits the nerve cells to multiply by a factor of 2 at least.

Development of the human psyche and the acquisition of reflective thought are revealed in prehistoric remains: tools and the mastering of fire, the establishment of habitations, cannibalism and burials, the birth of art, and eventually the neolithic ecological revolution. The latter led to the considerable technological advances enjoyed by present-day man. Among other consequences, however, it also entailed an increase in the number of wars and in man's increasing control of his own evolution.

Tools. Tools are the first manifestation of reflective thought found in the prehistoric records. The fabrication of a tool implies the visualization of its potential use in a certain task. Its shape represents the attempt to achieve an efficient result. Use of a tool necessitates a clear understanding in the brain, and is then effected in an ordered sequence of operations.

The choice of raw materials is equally important. Specialists in this field are of the opinion that the manufacture of the first tools resulted from numerous states of trial-and-error and that the respective technology was then handed down through the generations by teaching. This implies that there was a language beyond that of the noised made by chimpanzees. The production of tools, according to a standard model, is actually beyond the brain power of non-humans. However, Jane Goodall [28], has shown that chimpanzees are able to use twigs, after pulling off the leaves, to catch termites by sticking this rudimentary tool into the openings of termite mounds, although the chimpanzees are not yet able to establish a real industry by repeated production of these tools. Thus, tools are clear evidence of the human psyche and of imagination, intent, and a chain of complex successive operations. The oldest tools which may be clearly dated are found in about 2-million-year-old deposits and are contemporary with the oldest remains of *Homo habilis*.

Fire and Habitation. The mastering of fire and the establishment of dwellings made their appearance with *Homo erectus*. Traces of burned bones and ash from campfires are known from an approximately 600 000-year-old cave at Chou-Kou-Tien (China) and from an approximately 450 000-year-old site at Verteszöllös (Hungary). In addition to improving the quality of life by facilitating the cooking of food, providing lighting, and warmth during cold periods, and by offering protection against wild beasts, fire also led to a greater social cohesion of groups of humans. From this time inwards, we observe traces of habitations used as resting areas, flint workshops, and places of slaughter. However, it is not until the appearance of Neanderthal man *(Homo sapiens neandertalensis)* that we find around the fireplaces, hut-like structures of arrangements of mammoth bones, which were intended to act as a superstructure, covered with skins.

Burials and Man's Self-Discovery. The most significant observations pertaining to the development of the human psyche are represented by burials which appeared only with the Mousterian civilisation, which might be ascribed either to Neanderthal man or to modern man from the Middle East. The presence of burials shows that man had acquired a conscience and a concept of his own identity. The performance of burials in combination with accompanying offerings suggests that there was a belief in another lif and that there have been mythologies which have long since been forgotten. Death had clearly taken on a magic significance and the despair exhibited during burials might serve as proof of a desire for protection from the dead, who were thought to "survive" in the tomb or its vicinity. In order to avoid becoming molested by the dead, some peoples even went to the extreme of devouring them.

Art and Abstraction. The next phase corresponded to the establishment of a range of arts and painting by modern man, starting about 30 000 years ago. This prehistoric art signified the new mental aptitude for symbolic expression, and resulted from modern man's powers abstraction.

The Ecological Revolution. The worldwide ecological, technological, economic, and social revolution of the Neolithic coincided with the end of the last glaciation, about 10 000 years ago. The climatic changes affected both fauna and flora, and man had to change his methods of obtaining food. Instead of living by hunting and gathering, man started to cultivate plants and to keep animals. As a breeder and farmer utilizing local species, man became settled, establishing village communities of herdsmen and farmers, at first in the Middle East some 6000 years ago. This radical change in life style constituted the first stage of man developing an independence from the environment. By building up stocks, man was no longer dependent on gathering and hunting, which had become difficult due to the climatic fluctuations. This ecological revolution was accompanied by a technological revolution in developing storage facilities, especially in the form of ceramic containers, and by an industrial revolution about 4500 years ago which culminated in the discovery of metallurgical processes. This economic, and the ensuing social, revolution led to the establishment of markets which eventually turned into the first cities. During this period the first kingdoms were founded and a complex social hierarchy became established. One of the results of this revolution was the invention of writing which, beyond its practical everyday use, became the vehicle for religious thought and history, the memory of humanity. As stated by Francis Hours [29[this change in the way of life ended "the terrestrial paradise of the hunters and gatherers, to lead us into the hard world of labour". However this change did not take place everywhere and occurred at different rates in different places; indeed some isolated peoples continue to live as hunter-gatherers. Numerous others still live as herdsmen and farmers, whereas the Western civilisation experienced a fantastic technological breakthrough.

Man, the Most Anti-Ecological Animal in Nature. Along with the ecological revolution, another cultural phenomenon made its appearance, namely, war, which is supported by the development of arms. It results from the collective aggressivenes of two opposing groups. This aggressiveness is developed and continued by education and socioculture as the "major shame of humanity". Massive exterminations of humans has marked the history of mankind for the last 8000 years, and has introduced the new factor of "cultural selection".

Through technological advances man has achieved a much wider geographical distribution than has been possible for any other species. He became less and less dependent on the environment, which he was able to modify increasingly, and was able to move about at will, even reaching the moon. H. Tintant [30] has expressed this in the remark: "Man is the most anti-ecological animal on this planet".

Through the development of medicine, man started to escape some of the constraints of biological evolution and to control his own survival. This struggle against natural selection resulted in a profound reduction in infant mortality. By increasing the control of procreation man can modify his own evolution and may soon be able to select the sex of offspring, and possibly also their characters, through "genetic engineering".

Finally, man is the only species in this planet which is able to destroy itself along with the biosphere. This technological power burdens mankind with the huge responsibility of mastering his own evolution.

Liberty or the End of the Dictatorship of the Genes. The preceding paragraphs have shown that human evolution was executed by the same biological mechanisms as those of other vertebrates. The fundamental difference between man and the other species rests in one of the consequences of this biological evolution, the acquisition of reflective thought. This enabled man to modify the course of evolution by combining two innovations, sociocultural evolution and the concept of liberty. Sociocultural is handed from generation to generation by teaching and by written and now including televisual material; production of the latter is expanding rapidly in our Western world. In animals, behavior is controlled by information in the genetic programme of the species, and is thus partly inborn. In contrast to this, man's behaviour is controlled by social and cultural traditions and is thus acquired by experience. Except in the field of behaviour, cultural evolution does not relate to biological evolution. As a consequence it becomes able to influence the parameters controlling this biological evolution by the struggle against natural selection, by modifications of the environment, or by the introduction of new social factors of selection, like social hierarchy or wars. While biological evolution appears to us to be unpredictable due to the importance of random events, human evolution is free, liberated somehow from the dictatorship of the genes, and becoming instead their masters.

This immense responsibility demands the acquisition of a code of ethics which has to be respected on a global scale. This is the price humanity has to pay for its own future.

Is Man Continuing to Evolve?

This frequently posed question meets with a positive reply. The evolutionary trends which have characterized the human lineage so far are actually still operative. One of the most remarkable trends concerns the reduction in the masticatory system; this has been going on for at least 2 million years and still continues today, as any dentist will be able to tell you.

We can note the increasing frequency of the disappearance of the so-called wisdom teeth, the third molars which now do not emerge in about 40% of the population. Due to a lack of space in increasingly smaller jaws, these teeth have to be extracted in another part of the population. Since about a dozen years ago, dentists started to note that the jaws of children, at least in western countries, are becoming reduced in size. Furthermore, there is a reduction in the size and a change in the shape of the second pair of incisors, called "eye teeth". It is conceivable that these will eventually disappear completely. This trend, initiated some 2 million years ago, is still continuing and it is difficult to envisage where it will end, and whether modern eating habits will influence its course.

The general increase in body size, observed in developed countries, is not necessarily an evolutionary trend as it is tied to a qualitative and quantitative improvement in diet and environmental conditions. Differences in the size and length of the arms, for instance, exhibit a positive correlation with the mean temperature of the environment.

These trends may also become reversed with changes in local conditions. However, if the trend is maintained or becomes established worldwide, it could become an evolutionary feature.

A Scenario for Man in the Third Millennium

If the evolutionary trends observed in the human lineage were to continue, we could make an estimate of the shape which man may acquire in the future millennia. Worldwide, man would be larger with a rounded head and a larger cranial capacity. The most obvious points of divergence would concern the face. The jaws would become shorter, the number of teeth would be reduced by the disappearance of the wisdom and eye teeth, and the chin would grow and become increasingly pronounced.

However, all the data presented in this book show that the evolution of organisms results from the interaction of internal and external constraints together with accidental events, and that, as a consequence, it is unpredictable. Such a scenario therefore has not predictive value. furthermore, we have shown that cultural evolution has superimposed itself in biological evolution and has thereby acquired increasing importance, enabling it to completely unsettle biological evolution.

Are There Human Races?

At first sight, the human species exhibits a very strong variability and this was interpreted by the 19th century anthropologists in terms of racial groups. Each specialist recognized a number of races on the basis of skin colour, type of hair, cranial parameters, and a number of non-metrical, or even psychological, characteristics. These attempts at a classification were, in most cases, used for "racist" purposes, in other words to establish a hierarchical valuation of the defined groups. The applications of these wild notions for political means has, unfortunately, recently acquired too much importance to be ignored here.

Serious scientific studies by André Langaney [31] and Albert Jacquard [32] in the variation of skin colour have shown that the difference between two individuals of the same group may be much greater than between the average values of two groups belonging to distinct "races".

The results were confirmed by investigations on blood groups and variations in the associated systems (Rhesus factor or HLA). It has been shown scientifically that populations may be distinguished from each other not by whether or not they possess a certain gene, but by the fact that the relative frequency of these genes differs markedly.

Furthermore, there is often a trend to confound biological differences with cultural differences which, when they accumulate, tend to accentuate the apparent divergence between the groups.

Analysis of human genetic diversity has shown that if one tried to compare the difference between the three classic groupings (yellow, white, and black) with the difference between nations of one skin colour or with the difference between individuals of the same nation, the respective percentages would be 7, 8 and 85%. This shows that within human genetic diversity, the variability between individuals of the same group is the strongest at 85% and that differences between groups or nations do not together exceed 15%. The results of these studies illustrate that the groups are not sufficiently separate for a significant genetic difference between human populations to be observed and that, consequently, there is no justification for racial classifications.

This result may appear paradoxical to those impressed by the colour differences between people when encountering them daily in the streets, but it is, nevertheless, a well-established scientific fact.

All individuals of the human species, different as they may appear, are able to interbreed with each other, thereby contributing to a mixing or homogenization of the genetic programme. This process is increased by the development of means of transport for moving ourselves around the planet. R. Lewontin [33] has shown that "about 75% of the genetic programme known for man is identical to all humanity, whatever their geographical origin".

Although the genetic and morphological diversity of man may lead to a certain disparity or inequality between individuals, it still represents the great potential of mankind, as it undoubtedly contains, hidden in the genetic programme of certain individuals, new adaptive possibilities which have not yet been executed, but which may prove to be of great use to man in the future. It might even your be neighbour who carries this potential.

Can We Measure Human Intelligence?

This question frequently surfaces in debates on subjects such as the preceding one. Because of certain erroneous notions on the subject, both verbal and written, we shall dwell in the issue for a while.

Intelligence, a characteristic of human reflective thought, has not yet been precisely defined, as nobody knows by what criteria it should be measured. A multitude of associated aspects illustrate the complexity of the problem. Intelligent activities are more or less based on perception, instinct, and learning. Although intelligence is a fluid concept, numerous authors have attempted to measure it empirically. As a result, a number of intelligence tests have been introduced, amongst which the "intelligence quotient" (I.Q.) test is the most famous.

The earliest attempt at testing was carried out by the Frenchman Alfred Binet who tried to identify among children with schooling problems the cases of dyslexia which needed help. The original aim of the test was changed by Lewis N. Terman and it was then used to classify children according to a "scale of mental capacities" which was supposedly independent of education.

One of the most obvious conclusions drawn from these tests was that they actually evaluated the resistance of students to spending a long time in solving problems which made no apparent sense and had no practical use. The tests evaluated the attention of students and while this is indeed important for schooling, it has little to do with intelligence. According to R. Lewontin [33] one of the most remarkable results of the tests was that they showed that "it is worth more to be born rich than to be intellectually talented."

Investigations on the significance of the I.Q, of identical twins (with the same genetic make-up) were carried out on pairs which had been raised either separately or together. It was found that the I.Q. values of identical twins who had been educated together possessed a correlation coefficient of 0.85, whereas for separately educated twins the correlation was 0.75. The similarity between these results implies that there could be some genetic connection, while the difference underscores the influence of the family environment on I.Q.

The most regrettable aspect of the application of these tests comes from the recent revelation that the numerous papers published by Cyril Burt over the last 20 years, and which made him an authority on the subject, were based in false test procedures. The most elementary precaution would therefore be to deny these tests the apparent significance which they do not actually possess.

We may conclude that man is the result of ongoing biological evolution. However, this biological evolution has equipped man with a large brain capable of making numerous connections, and culminating in a new function, namely, reflective thought. This novel human capacity has started to unbalance the course of evolution. Through this ability, man was able to found and develop cultures, and cultural evolution, passed down from generation to generation, has superimposed itself on biological evolution and started to interfere with it.

By "extending" his hand through the development of tools, mans has multiplied his physical possibilities. As shown by offerings presented to the dead, he acquired a perception of his unique position on Earth and of his differences from the animal world. By extending recent technological developments to the animal, vegetable, and mineral worlds, man began to exploit them in order to no longer be at the mercy of the environment. Man can modify his environment and even takes it into space. He is increasingly able to control his own reproduction, life span, and even the genetic pool of his own species. And last, but by no means least, man has developed a potential for destruction which would enable him not only to destroy his own species, but also anything else living on Earth.

As man increasingly becomes master of his own evolution, he will be more and more subject to the constraints and random events of political history. This implies that, unfortunately, he is at the mercy of irresponsible and fanatical individuals who would render human evolution completely unpredictable. Man's survival should thus become a matter of concern for all humanity. This requires a universal conscience emanating from the establishment of a defined code of ethics that is respected worldwide.

Part V
The Future of the Theory of Evolution

Chapter 15

What Will Become of the Theory of Evolution in the Year 2000?

The formulation of the synthetic theory represented a significant-advance for evolutionary studies, which had previously been hampered by apparently irreconcilable contractions. Combining Mendelian laws with Darwinism, it joined the antitheses in a synthesis which was subsequently refined and completed. Today, its content may not be simply rejected or ignored, although there are those who think of it as little more than a corpse.

We would reiterate that the data of the synthetic theory are too important to be ignored. However, lest it become a dogma, the synthetic theory should be rejuvenated and incorporate all the new data which various evolutionary studies continue to produce. In the years to come, this will lead to a more wide-ranging, explicative system which, will however, preserve its essential synthetic characteristics. Whatever we choose to call it – theory of the year 2000, renewed synthetic theory, – the name is of no importance. It will not be built on the ruins of its predecessor but rather on the solid foundations of the same.

This theory cannot be reductionist, reducing everything only to the genetic determinism, as is proposed by the so-called non-Darwinian theory of evolution. Between the gene and an organism integrated in the population of a living species, too many processes and events take place which modify the original determinism encoded by the gene. Therefore, such a reductionist approach to the biological world is insufficient to explain the complexity of the present or past organic world.

In its present "orthodox" form, the synthetic theory has become outdated in an number of aspects. It does not recognize molecular genetics which has revealed important novelties, like the enhanced capacity for expression by genes, or the functional hierarchical organization of genetic material in which jumping genes may cause major evolutionary events such as speciation. This concept will make its own way, so let us wait. Information from molecular genetics, in combination with developmental genetics, illustrates that changes may be initiated by the economic use of genetic information: a gene or small group of genes will suffice to start a whole chain of events. Structural genes, acting in isolation or in groups, will enhance multiple regulatory commands, thereby increasing the change of the same modification being repeated.

The synthetic theory ignored the development of an organism for the reasons outlined above. Certain authors, such as Goldschmidt, Waddington, and de Beer, raised their voices against this shortcoming; however, their comments have been heeded only in recent years. It is now an established fact that no synthesis can ignore embryology and developmental genetics, which are intimately connected with genetic determinism in an operational system, termed epigenotype by Waddington.

The whole phase of processing of the genetic information offers a multitude of new possibilities for changes of evolutionary impact. However, one problem still remains: The channels of development should actually be stable, otherwise there would be no species, but at the same time modifiable, as otherwise the species would remain immutable. How could they be changed? We do not know yet, as so far no one has been able to change the development of one animal species into that of another. The solution of this key problem will open the way for experimental evolution.

The new approach of the genetics of development has already yielded valuable information. It has shown that small and far-reaching transformations may result from a reduced genetic determinism: certain activated or repressed genes, some epigenetic mechanisms acting in sequence, and a transformation is established. This leaves us with another problem: signals or mechanisms control the formation of a leg, they elaborate a shape, but how does this occur? This is another excellent field for future research.

The discovery of temporizing genes was another important step. They control timing, but how do the cells read them? Another problem! But let us look first at the already truly spectacular findings. These faithful timekeepers are sometimes disturbed by mutations. There is a certain change of events in the sequence of development which, as a result, becomes unbalanced. Some events may be added and others eliminated, both with potentially considerable consequences. These are the chronological shifts or heterochronies, to which, at present, increasing importance is assigned. Their action may actually explain how, from a simple genetic modification, changes in the appearance of the final animal may result (consider the axolotl), or there may be an abrupt switch from one type of construction to another at a bifurcation without intermediate forms (Oster and Alberch, Chap. 6, [7]). At least in some cases we are now relieved of the necessity of considering intermediate forms, often so difficult to find or even conceive of. They simply do not have to exist.

Evolution may perform jumps just as it may advance slowly and continuously. This is another concept which the modified synthesis will have to incorporate.

Let us next consider the problem of "liberty" in evolution. Can it execute anything; is anything possible with time? What are the roles of chance and determinism in its course? We have already discussed this at length. Throughout the life of individuals and populations, chance will intervene at the mutational level. However, chance is continuously counteracted and controlled by external and internal constraints. The only external constraints the synthetic theory had recognized was natural selection. Recent investigations reveal the intervention of early internal constraints as a manifestation of the genetic information throughout development. These constraints are of such importance that we could say, along with Gould and Lewontin, that by the very processes of its construction the organism largely controls its evolutionary destiny.

What then is the position of natural selection in the synthetic theory? It remains important and we should not forget that, so far, selection is the only established system of real directive forces. The discovery of internal constraints has taken away some of its overpowering importance. What will be the future of natural selection? We would stress, along with H. Tintant, that, whatever its importance, selection does not create but only organize. Any ambiguity, as is still encountered in some cases, should be definitive removed.

An essential feature of the synthetic theory is the inclusion of populations at all levels, from the genes to the organisms in their environment, and to the higher level of integration of the species itself. For the living species which we observe today in a limited section of time, we possess the excellent definiton formulated by E. Mayr. However, for palaeontologists and evolutionists it is necessary to consider species in the temporal dimension, to extrapolate from the static to a dynamic point of view, continuous in time and space. A coherent synthesis of evolution must stress the links developed so far between the zoological and palaeontological approaches to the species.

Sustained efforts since the start of the synthetic theory have made it possible to develop models for the formation of new species, the key process of evolution. These models were formulated within the framework and spirit of the synthetic theory and are part of its heritage, A critical review and re-evaluation of the various models would be most welcome, as some models appear to be just repetitions of others. We have to remember that we do not know anything about the genetics of speciation or the possible determining factors of these processes.

The question of integrational levels has been discussed repeatedly. To recapitulate: at any level, whether it entails genes, cells, tissues, organs, or organisms, new properties, which could not be predicted from the preceding level, may become established. This phenomenon excludes any idea of reductionism to the genetic level only, and must be taken into account by the new synthesis which should, in turn, be constructed of a hierarchical sequence of organizational levels.

Concerning the levels, we have arrived at the much debated question of the levels of evolutionary realization. Is evolution a unitary process which, as proposed by the synthetic theory, modifies species, causing them to change from one to another, and leading to the major body plans? Or does evolution work at different levels, with different mechanisms responsible for micro-, macro-, and megaevolution? Are these levels controlled by genetic determinism or by different selective processes, as Stanley has proposed for micro- and macroevolution? As we have seen, the genetic programme and development do not exhibit different functional types, but are able at the same time either only to modify a structure or to alter it completely. On this basis is there then any justification for distinguishing between at least two types of evolution? In principle: No! But, on the contrary, palaeontologists have found numerous examples of progressive change which could be explained by microevolution, and there are also numerous examples of radical rapid changes which do not fit into the microevolutionary model. It is true that palaeontological time is really condensed, creating impression of a process taking place rapidly, if not immediately. Nevertheless, there are cases of evolution which are truly rapid and radical; human evolution presents a particularly instructive example. Should we then consider these differences in rate, which appear to justify a distinction between micro- and macroevolution, as resulting from different stimuli? On this question the ink has scarcely had time to dry.

In conclusion: Who will formulate this new synthesis or, rather, which group of minds will elaborate it? This clarification appears necessary as it is unlikely that only one person will be able to undertake a task that involves many disciplines of science. It is no longer sufficient to consider only genetics, Darwinism, and palaeontology as the sole components. Already these are being joined by experimental and genetic embryology, biochemistry, ecology, and ethology, and, no doubt other, sciences have still to come. The task will be enormous!

In the eyes of those specializing in the theory of science, the theory of evolution does not obey the criteria which they have established for a definition of a "scientific theory". they consider it as something more metaphysical (K. Popper, Chap. 1 [3].) We should recall that these authors are only discussing, ad infinitum, the theories of physics and that they tend to shy away from the domain of biology or, if they look at it all, produce interpretations of little interest. Organic evolution is not a never-ending repetition of the same process. It is based on a succession of events which, by their very nature, are not repetitive. Furthermore, the theory of evolution cannot predict anything as it is constantly affected by chance. Unpredictability is part of its nature, although this factos must not be used to deny it the status of a scientific theory. To underline this, let us recall with Dobzhansky (Chap. 1 [6]) that "nothing in biology makes sense except in the light of evolution".

Epilogue: What Has Become of the History of the Giraffe and Its Neck?

We have pointed out that the history of the giraffe's neck served as an example to explain the theoretical concepts of transformism, proposed by J.-B. Lamarck, and of Darwin's natural selection. However, at that time it had not been substantiated by palaeontological data. Since then, the record has furnished a number of fossils which enable us to reconstruct the major trends in the evolution of the giraffides of Africa, Asia, and Europe [1, 2, 3] (Fig. 16.1; Appendix 16.1).

Within the giraffe group, from the oldest Tertiary forms to the present form, there is a general trend of increasing size and a lengthening of the neck and front limbs. This trend, however, is far from obligatory as is illustrated by the okapi, a smaller sibling species of the giraffe. This evolution took place by successive speciations, during which morphological modifications became established, and which undoubtedly resulted from changes in development. The phases of speciation were succeeded by phases of relative stability. During these, the species were subjected to fluctuations in size, and possibly also in skin pattern, resulting from pronounced intraspecies variability. This may be observed at present in Africa where the species is divided into a number of subspecies.

However, the palaeontological record shows that in the oldest deposits, the giraffe was represented by specimens which exceeded in size even the largest current giraffes. This is in contradiction to what we might expect from theoretical considerations on evolutionary trends, such as an apparent general increase in size. The evolution of the giraffes therefore appears to represent a particular case!

Such evolution fits neither into the concepts of Lamarck nor into those of Darwin. Instead it constitutes a positive test of the concepts which we have developed in our approach to the new evolutionary synthesis. And, furthermore, it appears to be an excellent illustration of a particular case of the punctuational model in which a species, once formed, remains stable.

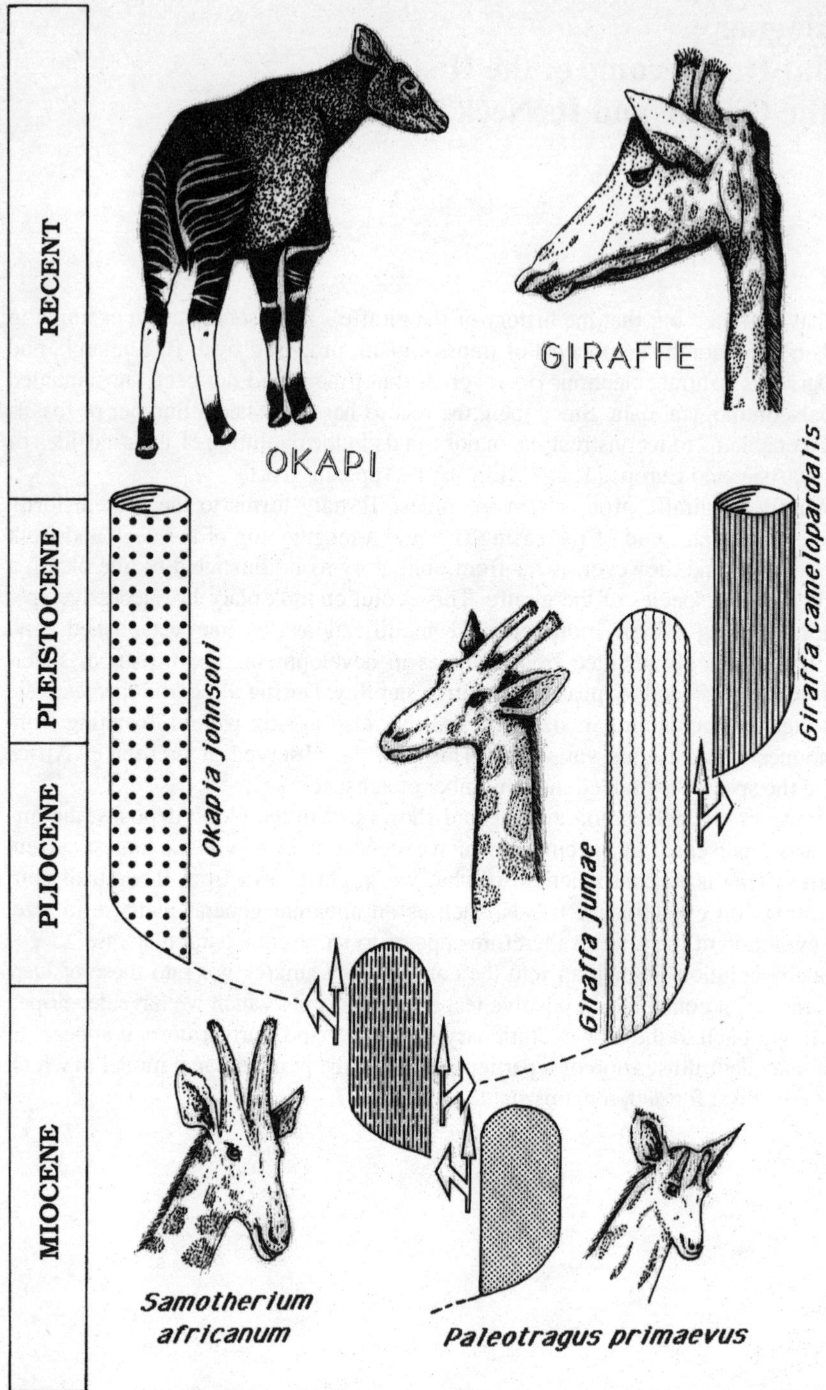

Fig.16.1. The neck of the giraffe: the present version.

Bibliography

The Theory of Evolution

1. Lamarck JB de (1809) Philosophie zoologique, 2 vol. Dentu, Paris (réédition Culture et Civilisation, Bruxelles, 1970)
2. Darwin C (1876) L'origine des espèces. Reinwald, Paris (réédition Maspero 1980)
3. Savage JM (1971) Evolution. Holt, London

Chapter 1

1. Whittaker RH (1969) New concepts of kingdoms of organisms. Science 163: 150–160
2. Devillers C, Mahé J (1980) Mécanismes de l'évolution animale. Masson, Paris
3. Popper K (1981) La quête inachevée. Calmann-Lévy, Paris
4. Popper K (1978) La logique de la découverte scientifique. Payot, Paris
5. Thuillier P (1987) La Science existe-t-elle? Le cas Pasteur. La Recherche 187: 506–511
6. Dobzhansky T (1973) Nothing in biology makes sense except in the light of evolution. Am Biol Teach 35: 125–129
7. Costa de Beauregard O (1982) Le paradigme de la mécanique quantique relativiste. Epistémologie de l'evolution. Annales Centre Etude Evolution Homme et Nature (CEHN), Paris, pp 35–43
8. Barreau H (1982) La dimension épistémologique de la recherche scientifique et technologique. epistémologie de l'évolution. Amales Centre Etude Evolution Homme et Nature (CEHN), Paris, pp 44–50
9. Gould SJ, Edlredge N (1977) Punctuated equilibria: the tempo and mode of evolution reconsidered. Palaeobiology, 3/2: 115–151
10. Delage Y, Hérouard E (1898) Traité de zoologie concrète, T VIII Les Procordés. Schleicher, Paris

Chapter 2

1. Lamarck JB de (1809) Philosophie zoologique, 2 vol. Dentu, Paris
2. Laurent G (1987) Paléontologie et évolution en France 1800–1860. De Cuvier-Lamarck à Darwin. Mémoires de la section d'histoire des sciences et des techniques, vol 4. CTHP, Paris, 553 pp

3. Darwin C (1859) On the origin of species by means of natural selection, or the preservation of favored races in the struggle for life. Harvard University Press, Belknap Press, Cambridge (Fac-similé, E Mayr, ed)
4. Mayr E (1982) The growth of biological thought. Harvard University Press, Cambridge
5. Malthus TR (1798) Essai sur le principe de population. Traduction E Vilquin sur la 1ère edn. 1980 PUF, Paris
6. Darwin C (1872) La descendance de l'homme. Traduction E Barbier sur la 3e édn anglaise (1881) (réédition Culture et Civilisation, Bruxelles)
7. Delage Y (1895) La structure du protoplasma et les théories sur l'hérédité. Reinwald, Paris
8. Mendel G (1866) Versuche über Pflanzen-Hybriden. Verh Naturforsch Brünn 4: 3–4
9. Orel Y, Armogathe JR (1985) Mendel. Belin, Paris
10. Bocquet C (1974) Précis de génétique formelle. PUF, Paris
11. Teissier G (1962) Transformisme d'aujourd'hui. Ann Biol 7–8: 359–374
12. Mayr E, Provine W (eds) (1980) The evolutionary synthesis, perspectives on the unification of biology. Harvard University Press, Cambridge, 487 pp
13. Ayala FK (1978) The mechanisms of evolution. Evol Sci Am 239: 48–61
14. Dobzhansky T (1937) Genetics and the origin of species. 3rd edn (1951). Columbia University Press, New York
15. Simpson GG (1944) Tempo and modes in evolution. Columbia Univesity Press, New York
16. Simpson GG (1950) Rhythmes et modalités de l'évolution. Traduction P de Saint-Seine. Albin Michel, Paris
17. Anxolabehere D, Periquet G (no date) Éléments de génétique des populations, vol 1 CDU, Paris
18. Devillers C (1985) Quelques remises en cause de la théorie synthétique de l'évolution. Ann Biol 24: 153–177
19. Wolter R, Grandjean D, Mateo R (1983) L'alimentation du chiot en croissance. Pratique médicale et chirurgicale de l'animal de compagnie 18 (3): 10–19
20. Simpson GG (1961) Horses. The Nat Hist Library, New York
21. Kettlewell HBD (1961) The phenomenon of industrial melanism in Lepidoptera. Ann. Rev Entomol 6: 245–262

Chapter 3

1. Gros F (1986) Les secrets du gène. O. Jacob, Paris
2. Bocquet C (1974) Précis de génétique formelle. PUF, Paris
3. Nei M (1987) Molecular evolutionary genetics. Columbia University Press, New York
4. Dobzhansky T, Boesiger E (1968) Essais sur l'évolution. Masson, Paris
5. Kimura M (1968) Evolutionary rate at the molecular level. Nature 217: 624–626
6. King JL, Jukes TH (1969) Non-darwinian evolution. Science 164: 788–798
7. L'Héritier P, Naefs Y, Teissier G (1936) Aptérisme des insectes et sélection naturelle. C.R. Acad. Sci, Paris 204: 907–910
8. Brunhes J (1985) Pourquoi certains insectes perdent leurs ailes? La Recherche 167: 830–833
9. Darlington J.Jr (1943) Carabidae of mountains and islands: data on the evolution of isolated faunas, and on atrophy of wings. Ecol Monogr 13: 37–61
10. Fisher RA (1958) The genetical theory of natural selection, 2nd édn. Dover, New York
11. Anxolabehere D, Periquet G (no date). Éléments de génétique des populations. CDU, Paris
12. Milkman R (ed) (1982) Perspectives on evolution. Sinauer, Sunderland, MA
13. Dobzhansky T (1977) Génétique du processus évolutif. Flammarion, Paris
14. Lin Chao K, Vargas C, Spear BB, Coxe C (1983) Transposable elements as mutator genes in evolution. Nature 303: 633–635
15. Davidson EG, Britten RJ (1973) Organization, transcription, and regulation in the animal genome. Quart Rev Biol 48: 565–613

16. Valentine JW, Campbell CA (1975) Genetic regulation and the fossil record. Am Sci 63: 673–680
17. Petit C, Zuckerkandl E (1976) Évolution. Génétique des populations. Évolution moléculaire. Hermann, Paris
18. Ayala F (1982) Biologie moléculaire et évolution. Masson, Paris
19. Ruffié J (1982) Traité du vivant. Fayard, Paris

Chapter 4

1. Haeckel E (1877) L'Anthropogénie ou histoire de l'évolution humaine. Traduction Ch Letourmeau. Reinwald, Paris
2. Haeckel E (1898) Natürliche Schöpfungsgeschichte, 9. Aufl. Reimer, Berlin
3. De Beer GR (1958) Embryos and ancestors. Clarendon Press, Oxford
4. Tintant H (1979) La récapitulation et ses exceptions. In: Bons J (ed) Aspects modernes des recherches sur l'évolution. Mém Trav Inst Montpellier 10: 17–27
5. Dommergues JL, David B, Marchand D (1986) Les relations ontogenèse-phylogenèse: applications paléontologiques. Géobios 19: 335–336
6. Gould SJ (1977) Ontogeny and phylogeny. Harvard University Press, Cambridge
7. Garstang W (1922) The theory of recapitulation: a critical re-statement of the biogenetic law. J Linn Soc Zool 35: 81–101
8. Baer KE von (1828) Entwicklungsgeschichte der Tiere: Beobachtung und Reflexion. Bornträger, Königsberg
9. Dalcq A (1935) L'organisation de l'oeuf chez les Chordés. Gauthier-Villars, Paris
10. Spemann H (1936) Experimentelle Beiträge zu einer Theorie der Entwicklung. Springer, Berlin
11. Willier BH, Weiss PA, Hamburger V (1955) Analysis of development. Saunders, Philadelphia
12. Gachelin G (1988) Le locus-t de la souris. La Recherche 196: 152–161
13. Raff RA (1989) Macroevolutionary changes in echinoid ontogeny: mechanisms and phylogenetic implications. In: David B, Dommergues JL, Chaline J, Laurin B (eds) Ontogenese et evolution. Geobios Mém Spéc 12: 313–321
14. Spierer P, Goldschmidt-Clermont M (1985) La génétique du développement de la mouche. La Recherche 154: 452–461
15. Garcia Bellido A, Lawrence P, Morata G (1979) Le développement cloisonné de l'animal. Pour la Science 23: 106–114
16. Morata G, Lawrence PA (1977) Homeotic genes, compartments and cell determination – *Drosophila.* Nature 265: 211–216
17. Raff RA, Kaufman EC (1983) Embryos, genes and evolution, Macmillan, New York
18. Rensch B (1959) Evolution above the species level. Methuen, London
19. Duméril A (1867) Métamorphoses des batraciens urodèles à branchies extérieures du Mexique dits axolotls, observées à la Ménagerie des Reptiles du Muséum d'Hitoire Naturelle. Ann Sci Nat Zool 7: 229–254
20. Kollar EJ, Fisher C (1980) Tooth induction in chick epithelium: expression of quiescent genes for enamel synthesis. Science 207: 993–995
21. Tompkins R (1978) Genetic control of axolotl metamorphosis. Am Zool 18: 313–319
22. Ambros V, Horvitz HR (1984) Heterochronic mutants of the nematode, *Caenorhabditis elegans.* Science 226: 409–416
23. Wake DB (1980) Evidence of heterochronic evolution: a nasal bone in the Olympic salamander, *Rhyacotriton olympicus* J Herpetol 14: 292–295
24. Roux W (1881) Der Kampf der Theile im Organismus. Engelmann, Leipzig
25. Spiegelman S (1945) Physiological competition as a regulatory mechanism in morphogenesis. Rev Biol 20(2): 121–146

26. Bretscher A, Tschumi P (1951) Gestufte Reduktion von chemisch behandelten *Xenopus* Beinen. Rev Suisse Zool 58: 391–398
27. Bretscher A (1947) Reduktion der Zehenzahl bei *Xenopus* – Larven nach lokaler Colchicin-behandlung. Rev Suisse Zool 54: 273–279
28. Alberch P, Gale EA (1983) Size dependence during the development of the amphibian foot. Colchicine-induced digital loss and reduction. J Embryol Exp Morphol 76: 177–197
29. Devillers C (1965) The role of morphogenesis in the origin of higher levels of organization. Syst Zool 14: 259–271
30. Hampé A (1958) Le développement du péroné dans les expériences sur la régulation des déficiences et des excédents dans la patte du poulet. J Embryol exp Morphol 6: 215–222
31. Gould SJ (1966) Allometry and size in ontogeny and phylogeny. Biol Rev, 41: 587–640
32. Devillers C, Mahé J, Ambroise D, Bauchot R, Chatelain E (1984) Allometric studies on the skull of living and fossil Equidae (Mammalia: Perissodactyla). J Vertebr Paleontol 4: 471–480
33. Gould SJ (1974) The origin and function of bizarre structures: antler size and skull size in the Irish Elk *Megaloceros giganteus*. Evolution 28: 191–220
34. McKinney ML, Schoch RM (1985) Titanothere allometry, heterochrony, and biomechanics: revision an evolutionary classic. Evolution 39: 1351–1363
35. Landauer W (194) The ancon or otter sheep. J Hered 40: 105–112
36. Hinchiffe JR, Johnson DR (1980) The development of the vertebrate limb. Clarendon Press, Oxford
37. Keibel F (1897–1909) Normentafeln zur Entwicklungsgeschichte der Wirbelthiere. Gustav Fischer, Jena
38. Gehring W (1985) Les molécules du développement: Pour la Science 98: 120–132
39. Gingerich PD (1981) Variation, sexual dimorphism, and social structure in the early Eocene horse *Hyracotherium* (Mammalia, Perissodactyla). Paleobiology 7: 443–455

Chapter 5

1. Devillers C (1981) Les vertébrés. CDU, Paris
2. Vuillemin S (1967) La respiration des Crustacés décapodes. Ann Biol 6 (1–2): 47–81
3. Johansen K (1970) Air breathing fishes. In: Hoar WS, Randall DJ (eds) Fish physiology, vol. 6. Academie Press, New York
4. Jacob F (1977) Évolution et bricolage. Le Monde 6 et 7/09 Evolution and tinkering. Science 196: 1161–1166
5. Gould SJ, Lewontin RC (1979) The spandrels of San Marco and the Panglossian paradigm: a critic of the adaptationist programme. Proc Roy Soc Lond B 205: 581–598
6. Dobzhansky T, Boesiger E (1968) Essais sur l'évolution. Masson, Paris
7. Ginsburg L (1984) Théories scientifiques et extinction des dinosaures. CR Acad Sci Paris 298: 317–320
8. Rubin CT, Lanyon LE (1983) Limb mechanics as a function of speed and gait; a study of functional strains in the radius and tibia of horse and dog. J exp Biol 101: 187–212

Chapter 6

1. Cuvier G (1830) Discours sur les révolutions de la surface du globe. (réédition C Bourgois, Paris, 1985)
2. Monod J (1970) Le hasard et la nécessité. Le Seuil, Paris
3. Thom R (1985) Pour ou contre Darwin? Science 1: 51–57

4. Wickramasinghe C (1982) L'astronomie: et si Darwin avait tort? Courrier Unesco (May 1982): 36–38
5. Gould SJ, Lewontin RC (1979) The spandrels of San Marco and the panglossian paradigm: a critic of the adaptationist programme. Proc Roy Soc Lond B 205: 581–598
6. Jacob F (1981) Le jeu des possibles. Fayard, Paris
7. Oster G, Alberch P (1982) Evolution and bifurcation of developmental programs. Evolution 36: 444–459
8. de Beer GR (1954) *Archaeopteryx* and evolution. Adv Sci 42: 1–11
9. Stanley SM (1979) Macroevolution. Pattern and process. Freeman, San Francisco
10. Waddington CH (1957) The strategy of the genes. Allen and Unwin, London
11. Goldschmidt R (1940) The material basis of evolution. Yale University Press, New Haven
12. Jacob F (1970) La logique du vivant. Gallimard, Paris
13. Gould SJ (1980) Is a new and general theory of evolution emerging? Paleobiology 6: 119–130

Chapter 7

1. Mayr E (1974) Populations, espèces et évolution. Hermann, Paris
2. Popper K (1954) Misère de l'historicisme. Plon, Paris
3. Bonhomme F, Thaler L (1988) L'évolution de la souris. La Recherche 199: 606–616
4. Chaline J (1987) Paléontologie des Vertébrés. Dunod, Paris
5. Rousseau D-D (1985) Structures des populations quaternaires de *Pupilla muscorum* (Gastropode) en Europe du Nord. Relations avec leurs environnements. Thèse 3e Cycle, Centre des Sciences de la Terre, Dijon
6. Villard F (1987) La variation morphologique chez *Vallonica costata* (Gastropoda, Pulmonata) en relation avec le milieu. Thèse EPHE, EPHE, Univ Dijon
7. Darwin C (1859) On the origin of species by means of natural selection. Harvard University Press, Belknap Press, Cambridge (Fac-similé, E Mayr, ed)
8. Mayr E (1982) Processes of speciation In: Mechanisms of speciation. Alan R Liss, New York pp 1–9
9. Carson H (1982) Speciation as a major reorganization of polygenic balances In: Mechanisms of speciation. Alan R Liss, New York, pp 411–433
10. Powell JR (1978) The founder-flush speciation theory: an experimental approach. Evolution 32(3): 465–474
11. de Vries H (1906) Species and varieties, their origin by mutations. Open Court, Chicago
12. Bateson W (1894) Material for the study of variation. MacMillan, New York
13. Goldschmidt R (1940) The material basis of evolution. Yale University Press, New Haven
14. Schindewolf O (1950) Grundfragen der Paläontologie. Schweizerbart, Stuttgart
15. Vandel A (1968) La genèse du vivant. Masson, Paris
16. Grassé PP (1978) Biologie moléculaire, mutagenése et évolution. Masson, Paris
17. Chaline J (1984) Le concept d'évolution polyphasée et ses implications. Gébios 17(6): 783–795
18. Ticehurst CB (1938) A systematic review of the genus *Phylloscopus*. British Museum, London

Chapter 8

1. Reeves H (1980) Patience dans l'azur. Le Seuil, Paris
2. Buvet R (1974) L'origine des êtres vivants et des processus biologiques. Masson, Paris

3. Miller SL (1986) Current status of the prebiotic synthesis of small molecules. Chem Scr 26B: 5–11
4. Dickerson RF (1978) Chemical evolution and the origin of life. Sci Am 239: 62-78
5. Valentine JW (1977) Cosmic evolution and the origin of life. In: Dobzhansky T, Ayala FS, Stebbins GL, Valentine JW (eds) Evolution. Freeman, San Francisco
6. Raulin F (1987) Formation et évolution des molécules organiques sur la Terre primitive. La Terre et l'origine de la vie. Colloq Eur CIEEIST, Université de Paris Sud (polycopié)
7. Barghorn ES (1971) The oldest fossils. Sci Am 224: 30–42
8. Schopf WJ (1978) The evolution of the earliest cells. Sci Am 239: 84–102
9. Schopf JW, Oehler DZ (1976) How old are the eukaryotes? Science 193: 47–49
10. Glaesner MF, Wade M (1966) The Late Praecambrian fossils of Ediacara, South Australia. Paleontology 9: 599–628
11. Odin GS (1982) The Phanerozoic time scale revisited. Episodes 3: 3–9

Chapter 9

1. Chaline J, Thaler L (1977) Le problème de l'espèce chez les rongeurs. Approche biologique et approche paléontologique. Les problèmes de l'espèce dans le règne animal. Soc Zool Fr 11: 359–374
2. Thaler L (1983) Image paléontologique et contenu biologique des lignées évolutives. In: Chaline J (ed) Modalités, rythmes et mécanismes de l'évolution biologique: gradualisme phylétique ou équilibres ponctués? Colloq Int Dijon 330. CNRS, Paris, pp 327–336
3. Gingerich P (1979) The stratophenetic approach to phylogeny reconstruction in vertebrate paleontology. In: Cracraft J, Eldredge N (eds) Phylognetic analysis and paleontology. Columbia University Press, New York, pp 41–77
4. Simpson GG (1953) The major features of evolution. Columbia University Press, New York
5. Hennig W (1950) Grundzüge einer Theorie der phylogenetischen Systematik. Deutscher Zentralverlag, Berlin
6. Mayr E (1974) Populations, espèces et évolution. Hermann, Paris
7. Chaline J (1984) Le concept d'évolution polyphasée et ses implications. Geobios 17(6): 783–795
8. Chaline J (1987) Arvicolid data (Arvicolidae, Rodentia) and evolutionary concepts. Evol Biol 21: 237–310
9. Gall JC (1971) Faunes et paysages des grès à Voltzia du Nord des Vosges. Essai paléoécologique sur le Bundsandstein supérieur. Mem Serv Carte Géol Alsace-Lorraine, Strasbourg 34: 1–318
10. Tintant H (1984) La stase des espèces est-elle un fait général? Le cas des céphalopodes fossiles. 109 e Congr Soc savantes, Dijon 1: 395–403
11. Laurin B (1983) Un test du bottleneck effect chez les Rhynchonelles jurassiques (Brachiopodes). In: Chaline J (ed) Modalités, rythmes et mécanismes de l'évolution biologique: gradualisme phylétque ou équilibres ponctués? Colloq Int Dijon 330. CNRS, Paris, pp 155–164
12. White M (1978) Modes of speciation. Freeman, San Francisco
13. Ozawa T (1975) Evolution of *Lepidolina multiseptata* (Permian foraminifer, East Asia). Mem Fac Sci Kyushu Univ Ser D Geol (23): 117–164
14. Kellogg DE (1975) The role of phyletic change in the evolution of *Pseudocubus vema* (Radiolaria). Paleobiology 1: 359–370
15. Kellogg DE (1983) Phenology of morphologic change in radiolarian lineages from deep-sea cores: implications for macroevolution. Paleobiology 9: 355–362
16. Tintant H (1963) Les Kosmocératidés du Callovien inférieur et moyen d'Europe occidentale. PUF, Paris
17. Tintant H (1983) Cent ans après Darwin, continuité ou discontinuité dans l'évolution. In: Chaline J (ed) Colloq Int 330. CNRS, Paris, pp 25–37

18. Gabilly J (1976) Evolution et systématique des Phymatocératinés et des Grammocératinés (Hildocératidés, Ammonitina) de la région de Thouars, stratotype du Toarcien. Mém Soc Géol Fr 124: 1–196
19. Dommergues JL (1984) L'évolution des Ammonitina au Lias moyen (Carixien, Domérien basal) en Europe occidentale. Thèse Univ, C Bernard, Lyon
20. Cariou E (1984) Les Reineckeidés (Ammonitina, Callovien) de la Téthys occidentale: systématique, dimorphisme et évolution. Etude à partir des gisements du centre ouest de la France. Thèse Univ Poitiers, Doc Lab Lyon 8
21. Marchand D (1986) L'évolution des Cardiocératinés d'Europe occidentale dans leur contexte paléogéographique (Callovien supérieur-Oxfordien moyen). Thèse Univ Bourgogne, Dijon
22. Hantzpergue P (1987) Les ammonites kimméridgiennes du haut-fond d'Europe occidentale (Perisphinctidae, Aulacostephanacae, Aspidoceratidae). Thèse Univ Poitiers
23. Gingerich P (1976) Paleontology and phylogeny: patterns of evolution at the species level in early tertiary mammals. Am J Sci 276: 1–28
24. Gingerich P (1985) Species in the fossil record: concepts, trends, and transitions. Paleobiology 11(1): 27–41
25. Heinrich WD (1982) Ein Evolutionstrend bei *Arvicola* (Rodentia, mammalia) und seine Bedeutung für die Biostratigraphie im Pleistozän Europas. Wiss Z Humboldt Univ Berl Math-naturwiss Reihe 31(3): 155–160
26. Chaline J, Laurin B (1986) Phyletic gradualism in a European plio-Pleistocene *Mimomys* lineage (Atvicolidae, Rodentia). Paleobiology 12(2): 203–216
27. Chaline J, Sevilla P (1990) Phyletic gradualism and developmental heterochronies in a European Plio-Pleistocene *Mimomys* lineage (Arvicolidae, Rodentia) Int Colloq Evolution, phylogeny and biostratigraphy of arvicolids. Czechoslovakia Geol Survey Praha 85–98
28. Martin LM (1984) Phyletic trends and evolutionary rates. In: Genoways H, Dawson M (eds) Contributions in Quaternary Vertebrate paleontology. Carnegie Mus Nat Hist Spec Publ 8: 526–538
29. Rousseau D-D (1985) Structures des populations quaternaires de *Pupilla muscorum* (Gastropode) en Europe du Nord. Relations avec leurs environnements. Thèse 3e Cycle, Centre des Sciences de la Terre, Dijon
30. Williamson PG (1981) Paleontological documentation of speciation in Cenozoic molluscs from Turkana Basin. Nature 293 (5832): 437–443
31. Freyer G, Greenwood PH, Peake JF (1983) Punctuated equilibria, morphological stasis and the paleontological documentation of speciation: a biological appraisal of a case history in an African lake. J Linn Soc 20: 195–205
32. Rey J, Fontaine P, Jimenez M-C (1986) Relations entre changements des paléomilieux et fluctuations des caractères évolutifs chez les foraminifères benthiques éocènes. Bull Centr Rech Explor-Prod Elf-Aquitaine 10(2): 369–382
33. Dommergues JL, David B, Marchand D (1986) Les relations ontogenèse – phylogenèse: applications paléontologiques. Géobios 19(3): 335–356
34. Alberch P, Alberch J (1981) Heterochronic mechanisms of morphological diversification and evolutionary change in the neotropical salamander *Bolitoglossa occidentalis* J Morphol 167: 249–264
35. Ricqles A, Taquet P (1982) La faune de vertébrés du Permien supérieur du Niger. 1. Le Captorhinomorphe *Moradiscus grandis* (Reptilia, Cotylosauria) – le crâne –. Ann Paléontol 68(1): 33–103
36. Goujet D (1981) Systematique et phylogenie. Encyclopedia Universalis, pp 356–357
37. Devillers C, Chaline J, Laurin B (1990) Plaidoyer pour une embryologie évolutive. La Recherche 21: 802–809
38. Alberch P, Gould SJ, Oster G F, Wake DB (1979) Size and shape in ontogeny and phylogeny. Paleobiology 5: 196–317
39. McNamara KJ (1986) A guide to the nomenclature of heterochrony. J Paleontol 60: 4–13

Chapter 10

1. Vandel A (1968) La genèse du vivant. Masson, Paris
2. Grassé PP (1978) Biologie moléculaire, mutagenèse et évolution. Masson, Paris
3. Goldschmidt R (1940) The material basis of evolution. Yale University Press, New Haven
4. Schindewolf O (1950) Grundfragen der Paläontologie. Schweizerbart, Stuttgart
5. Garstang W (1928) Preliminary note on a new theory of the phylogeny of the Chordata. Q J Microsc Sci 72: 51–62
6. Devillers C (1981) La genèse des mammifères. La Recherche 122(12): 580–589
7. Crompton AW (1958) The cranial morphology of a new genus and species of Ictidosaurian. Proc Zool Soc Lond 130: 183–215
8. Valentine JW (1980) L'origine des grands groupes d'animaux. La Recherche 112: 666–674
9. Martin LM (1984) Phyletic trends and evolutionary rates. In: Genoways H, Dawson M (eds) Contributions in Quaternary vertebrate paleontology. Carnegie Mus Nat Hist Spec Publ 8: 526–538
10. Gould SJ, Eldredge N (1977) Punctuated equilibria: the tempo and mode of evolution reconsidered. Paleobiology 3(2): 115–151
11. Stanley SM (1979) Macroevolution, pattern and process. Freeman, San Francisco
12. Vrba E (1980) Evolution, species and fossils: how does life evolve. S Afri J Sci 76(2): 61–84
13. Legendre S (1988) Les communautés de mammifères du Paléogène (Eocène supérieur et Oligocène) d'Europe occidentale: structures, milieux et évolution. Thèse, Univ Montpellier
14. Tintant H (1985) L'évolution des céphalopodes: gradualisme ou ponctualisme? La Vie des Sciences. C R Acad Sci Paris Ser Gén 2(5): 409–427
15. Dommergues JL (1984) L'évolution des Ammonitina au Lias moyen (Carixien, Domérien basal) en Europe occidentale. Thèse Univ, C Bernard, Lyon
16. Tintant H (1984) La stase des espèces est-elle un fait général? Le cas des céphalopodes fossiles. 109e Congr nat Soc savantes, Dijon. Sciences 1: 395–403
17. Romer AS, Price LI (1940) Review of the Pelycosauria. Geol Soc Am Spec Pap 28: 1–538
18. Parrington FR (1933) On the cynodont reptile *Thrinaxodon liorhinus* Seely. Ann Mag nat Hist 11: 16–24
19. de Beer GR (1937) The development of the vertebrate skull. Oxford Univesity Press, Oxford

Chapter 11

1. Tintant H (1983) Cent ans après Darwin, continuité ou discontinuité dans l'évolution? In: Chaline J (ed) Modalités, rythmes et mécanismes de l'évolution biologiqe. Colloq Int CNRS Dijon 330. CNRS, Paris, pp 25–37
2. Cuvier G (1812) Recherches sur les ossemens fossiles des quadrupèdes. Deterville, Paris
3 Eldredge N, Gould SJ (1972) Punctuated equilibria: an alternative to phyletic gradualism. In: Schopf T (ed). Models in paleobiology, Freeman, Cooper, San Francisco, pp 82–115
4. Gould SJ, Eldredge N (1977) Punctuated equilibria: the tempo and mode of evolution reconsidered. Paleobiology 3(2): 115–151
5. Simpson GG (1944) Tempo and mode in evolution. Columbia University Press, New York
6. Simpson GG (1953) The major features of evolution. Columbia University Press, New York
7. Huxley J (1942) Evolution, the modern synthesis. Harper, New York
8. Gould SJ (1983) Dix huit points au sujet des équilibres ponctués. In: Chaline J (ed) Modalités, rythmes et mécanismes de l'évolution biologique. Colloq Int CNRS Dijon 330. CNRS, Paris, pp 39–41
9. Eldredge N (1971) The allopatric model and phylogeny in Paleozoic invertebrates. Evolution 25: 156–167
10. Gould SJ (1969) An evolutionary microcosm: Pleistocene and recent history of the land snail *P. (Poecilozonites)* in Bermuda. Bull Mus Comp Zool 138(7): 407–532

11. Gingerich P (1976) Paleontology and phylogeny: patterns of evolution at the species level in early Tertiary mammals. Am J Sci 276: 1–28
12. Chaline J (1987) Arvicolid data (Arvicolidae, Rodentia) and evolutionary concepts. Evol Biol 21: 237–310
13. Erwin DH, Valentine JW, Sepkoski JJ (1987) A comparative study of diversification events: the Early Paleozoic versus the Mesozoic. Evolution 41(6): 1177–1186
14. Hallam A (1987) End Cretaceous mass extinction event: argument for terrestrial causation. Science 238: 1237–1242
15. Alvarez LW, Alvarez W, Asaro F, Michel HV (1980) Extraterrestrial cause for the Cretaceous-Tertiary extinction. Science 208: 1095–1108
16. Courtillot V, Besse J, Vandamme D, Montigny R, Jaeger J.J Capetta H (1986) Deccan flood basalts and the Cretaceous/Tertiary boundary. Earth Plan Sci Letters 80: 361–374
17. Martin LD (1981) Tertiary Extinction cycles and the Pliocene-Pleistocene Boundary. Institute for Tertiary-Ouaternary Studies-TER-QUA Symp Series 1: 33–40
18. Graham RW (1985) Response of mammalian Communities to Environmental Changes During the Late Quanternary. In: Diamond J, Case TJ (eds) Community Ecology. Harper & Row New York: 300–313
19. Chaline J (1984) Le concept d'évolution polyphasée et ses implicatons. Geobios 17(6): 783–795
20. Chaline J, Brunet Lecomte P (1990) Modélisation de l'évolution. C R Acad Sci Paris 311 (II): 10331–1036

Chapter 12

1. Grassé P-P (1971) Toi ce petit dieu. Essai sur l'histoire naturelle de l'homme. Albin Michel, Paris
2. Delsol M (1985) Cause, loi, hasard en biologie. Vrin, Paris et Institut Interdisciplinaire d'Etudes Epistémologiques de Lyon
3. Tintant H (1983) La paléontologie explique-t-elle l'évolution? In: Barreau H (ed) L'explication dans les sciences de la vie. CNRS, Paris, pp 161–196
4. Tintant H (1986) La loi et l'événement. Deux aspects complémentaires des Sciences de la Terre. Bull Soc Géol Fr 8(2,1): 185–190
5. Hofker J (1959) Orthogenesen von Foraminiferen. Neues jahrb Geol Palaeontol Abh 108: 239–259
6. Saint-Seine P de (1948) Les fossiles au rendez-vous des calculs. Etudes: 193–205
7. Depéret C (1907) Les transformations du monde animal. Flammarion, Paris
8. Newell ND (1967) Revolutions in the history of life. Spec Pap Geol Sec Am 89: 63–91
9. Dollo L (1893) Les lois d'évolution. Bull Soc Belge Géol 7: 164–166
10. Gould SJ (1970) Dollo on Dollo's law: irreversibility and the status of evolutionary laws. J Hist Biol Harvard College 3(2): 189–212
11. Gingerich P (1977) Pattern of Evolution in the mammalian fossil record. In: Hallam A (ed) Pattern of evolution. Elsevier, Amsterdam, pp 469–500
12. Cournot A (1922) Traité sur l'enchaînement des idées fondamentales dans les sciences et dans l'histoire. Hachette, Paris
13. Monod J (1970) Le hasard et la nécessité. Le Seuil, Paris
14. L'Héritier P (1954) Traité de génétique. PUF, Paris
15. Chaline J, Laurin B (1984) Le rôle du climat dans l'évolution graduelle de la lignée *Mimomys occitanus-ostramosensis* (Arvicolidae, Rodentia) au Pliocène supérieur. Géobios 8: 323–331

Chapter 13

1. White MJD (1978) Modes of speciation. WH Freeman, San Francisco
2. Dobzhansky T, Ayala F, Stebbins GL, Valentine JW (1977) Evolution. WH Freeman, San Francisco

Chapter 14

1. Grelot P (1986) La Bible et les origines de l'homme. In: Chaline J (ed) L'homme dans l'évolution de l'univers. dossier Histoire et Archéologie Dijon 114: 88–96
2. Darwin C (1874) De la descendance de l'homme en liaison avec la sélection sexuelle. Traduction E. Barbier sur la 3ème édition anglaise (1881). Ré-édition: Ed. Complexes, Bruxelles, 2 vols
3. Collet JY (1988) La planète des primates. Les petits des grands singes. Terre sauvage. Nuit et Jour: 6–12
4. King MC, Wilson AC (1975) Evolution at two levels in human and chimpanzees. Science 188: 201–203
5. Miyamoto MM, Koop BF, Slightom JL, Goodman M, Tennant MR (1988) Molecular systematics of higher primates: genealogical relations and classification. Proc Natl Acad Sci USA 85: 7627–7631
6. Grouchy J DE, Turleau C, Roubin M, Klein M (1972) Evolutions caryotypiques de l'homme et du chimpanzé. Etude comparative des topographies de bandes après dénaturation ménagée. Ann Génét 15: 79–84
7. Dutrillaux B, Couturier J (1986) Principes de l'analyse chromosomique appliquée à la phylogénie: l'exemple des Pongidae et des Hominidae. Mammalia 50: 22–37
8. Stanyon R, Chiarelli B (1982) Phylogeny of the Hominoidea: the chromosome evidence. J Hum Evol 11: 493–504
9. Chaline J, Dutrillaux B, Couturier J, Durand A, Marchand D (1991) Un modèle chromosomique et paléobiogéographique d'évolution des primates supérieurs. Géobios 24(1): 105–110
10. Bolk R (1926) Das Problem der Menschwerdung. Gustav Fischer, Jena
11. Chaline J, Marchand D, Berge C (1986) L'évolution de l'homme: un modèle gradualiste ou ponctualiste? Bull Soc R Belg Anthropol Préhist 97: 77–97
12. Dambricourt-Malassé A (1988) Hominisation et foetalisation (Bolk, 1926). C R Acad Sci Paris 307, II(2): 199–204
13. Deshayes MJ, Dambricourt-Malassé A (1990) Analyse des différents types architecturaux cranio-faciaux par l'approche ontogénétique de l'hominisation. Rev Stomatol Chir Maxillofac 91(4): 249–258
14. Simons E, Pilbeam D (1978) Ramapithecus (Hominidae). In: Maglio Cooke (eds) Evolution of African mammals, Vol 9. Harvard University press, Cambridge, pp 147–153
15. Simons E (1961) The phyletic position of *Ramapithecus*. Postilla 57: 1–9
16. De Bonis L (1980) Les primates humanoides du Miocène et le problème du Ramapithèque. In: Ferembach D (ed) Les processus de l'hominisation. CNRS Paris 599: 49–53
17. Lipson S, Pilbeam D (1982) *Ramapithecus* and hominoid evolution. J Hum Evol 11: 545–548
18. De Bonis L, Bouvrain G, Geraads D, Koufos G (1990) New hominid skull material from the late Miocene of Macedonia in northern Greece. Nature 345: 712–714
19. Andrews P (1990) Lining up the ancestors. Nature 345: 664–665
20. Hurzeler J (1968) Signification de l'Oréopithèque dans la phylogénie humaine. Triangle 4–5: 164–174
21. Bishop WW, Pickford M (1975) Geology, fauna and paleoenvironments of the Ngorora Formation, Kenya, Rift Valley. Nature 254: 185–192

22. Coppens Y (1986) Evolution de l'homme. CRAcad Sci Paris 3(3): 227–243
23. McNamara KJ (1990) The role of heterochrony in evolutionary trends. In: McNamara KJ (ed) Evolutionary trends: pp 59–74
24. Laitman JT (1986) L'origine du langage articulé. La Recherche 181 (17): 1164–1173
25. Chaline J, Brunet-Lecomte P (1990) Modélisation des modalités de l'évolution. C R Acad Sci Paris 311(II): 1031–1036
26. Shea BT (1984) An allometric perspective on the morphological and evolutionary relationships between pygmy *(Pan paniscus)* or common *(Pan troglodytes)* chimpanzees. In: Susman RL (ed) The pygmy chimpanzee. Evolutionary biology and behavior. Plenum Press New York, 89–130
27. Mayr E (1974) Populations, espèces et évolution. Hermann, Paris
28. Goodall J (1971) Les chimapnzés et moi. Stock, Paris
29. Hours F (1987) La révolution néolithique. In: Chaline J (ed) L'homme dans l'évolution de l'univers. Dossiers Histoire et Archéologie 114: 88–96
30. Tintant H (1987) Originalité de l'homme. In: Chaline J (ed) L'homme dans l'évolution de l'univers. Dossiers Histoire et Archéologie 114: 71–73
31. Langaney A (1977) La quadrature des races. Génétique et anthropologie. Sci Vie: 83–127
32. Jacquard A (1978) Eloge de la différence. Points Science Seuil, Paris
33. Lewontin R (1984) La diversité des hommes. L'inné, l'acquis et la génétique. Pour la Science, Belin
34. Mein P (1983) Histoire paléontologique des primates supérieurs jusqu'à l'apparition du genre *Homo*. Cahiers Inst Catholique Lyon 9: 31–43
35. Berge C, Orban-Segebarth R, Schmidt P (1984) Obstetrical interpretation of the Australopithecine pelvic cavity. J Human Evol 13: 573–587
36. Howell FC (1972) Pliocene/Pleistocene Hominidae in eastern Africa: absolute and relatives ages. In: Bishop W, Miller J (eds) Calibration of hominid evolution. Scottish Acad Press, Edinburg, pp 331–368
37. Walker A, Leakey R (1978) Les hominidés du Turkana oriental. Pour la Science 48–65

Epilogue

1. Thenius E (1982) Grundzüge der Verbreitungsgeschichte der Säugetiere. Fischer, Jena
2. Churcher CS (1978) Giraffidae. In: Maglio V, Cooke HB (eds). Evolution of African mammals. Harvard Press, Cambridge, pp 509–535
3. Geraads D (1986) Remarques sur la systématique et la phylogénie des Giraffidae (Artiodactyla, Mammalia). Geobios 19(4): 465–477
4. Murray J (1988) Les taches du léopard. Pour la Science 127: 78–86

Glossary

Acceleration. Increase in the rate of morphological development of a descendant organism, leading to an adult of the same size as the ancestor but with a more advanced or "hyperadult" morphology

Accretion. Type of growth of ammonite and nautilus shells through addition of chambers

Aerobic, Anaerobic (Respiration). In order to grow cells require a supply of energy from respiration. This may be aerobic, as found in bacteria and unicellar and multicellar organisms, or anaerobic, as found in certain bacteria; the type of respiration depends on whether free oxygen is available (aerobic) or not (anaerobic). In aerobic respiration a molecule rich in energy, like glucose, is oxidized in a complex of reactions (respiratory chain) to form carbon dioxide and water. Energy is distributed throughout a molecule like ATP (adenosine triphosphate) in small quantities, the so-called energy-rich bonds. These molecules then supply the energy necessary for the synthesizing reactions. Anaerobic respiration was obligatory for the first organisms, formed at a time when there was no dissolved oxygen present in the hydrosphere. This respiratory process also occurs in certain present bacteria. The energy donor here is a mineral compound like sulphate, and the respiratory chain leads to the production of water and hydrogen sulphide (sulphur bacteria).

Allele. One of several forms of the same gene (wild allele, mutant allele)

Allopatric. Two populations are called allopatric when their areas of distribution do not overlap (see also sympatric, parapatric)

Amphioxus. See Cephalochordata

Anagenesis. Progressive evolution within a lineage. In the synthetic theory anagenesis was considered to be an evolutionary mode. In our synthesis it is considered as a consequence of speciation modalities

Analogy/homology. Two analogous structures fulfil the same function, without having the same constructional principle, e.g. wings of *Drosophila* and birds. Two structures A and B may be called homologous when, amongst living forms, the development of B may be deduced from A through a number of modifications. The two structures exhibit the same fundamental organization, although the function itself is not taken into consideration, e.g. the articular bone of the reptilian jaw and the malleus of the middle ear of the mammals are homologous (Chap. 10). For fossil forms which are separated from each other by time, homology between A and B may be established either by tracing the transformation from A to B through successive intermediate stages in the lineage, or by showing that A and B still show the same arrangement with regard to adjoining structures (principle of connections by E. Geoffroy Saint-Hilaire). The presence of homologies allows the recognition of common origins and thereby the reconstruction of evolutionary lineages

Annelids. In the annelid (segmented worm) phylum, three types are recognized, namely, the bristle worm, the earthworm, and the leeach. In the latter, the coelom of the adult is filled by a proliferation of cells. Annelids live in fresh and marine waters and on land. The group originated from the subdivision of the ancient "worm" phylum which included other, rather different types of construction: platyhelminthes (flatworms), nematodes (round worms), etc.

Antigen/Antibody. An antigen is a molecule, virus, bacterium, or cell which, when introduced into a body, produces an immune defence reaction. The antibody formed by the immune system of the organism attaches itself to the antigen, inactivating or destroying it through the action of special cells (macrophages)

Arthropods. Numerically the most important animal phylum on Earth, with about 1 million species. It may be divided into several subphyla: arachnids (spiders), myriapods (millipedes) crustaceans, insects (the most numerous group), ... and the trilobites of the Palaeozoic. Their epidermis secretes a cuticle made up of chitin (a cellulose-like substance) which may remain flexible between individual segments or become hardened by "tanning" (as in the Coleoptera) or by calcification, as in the carapace of the lobster

Ascidiacea. Sea squirts; see also Tunicata

Atmosphere. The modern atmosphere is oxidizing, containing some 21% of oxygen which tends to combine spontaneously with a number of molecules. During the early history of the Earth the atmosphere was devoid of oxygen and was therefore reducing. It consisted of carbon dioxide, water vapor, nitrogen, ammonia, hydrogen sulphide (?), methane (?), etc. Its composition is still much debated

Belemnites. Mesozoic cephalopod molluscs comparable to the recent cuttlefish or squids. As fossils they are usually only preserved by their rostrum, the most durable part of their internal shell

Benthic. An aquatic organism living on the bottom of a body of water

Brachiopods. Marine organisms living in a bivalve shell. Comparatively rare at present, they were highly diversified during the Palaeozoic and Mesozoic

Bryozoans. A group of protostomes living in colonies with each individual occupying its own chamber; found in marine and fresh waters

Canalization. Term introduced by Waddington expressing the fact that developmental pathways lead to a well-defined organism, despite disturbances in the genetic instructions or environmental factors (also referred to as homeostasis of development)

Cell. In a prokaryotic cell, such as in bacteria and cyanobacteria (blue-green algae), the genetic material, consisting of single DNa molecules, is in direct contact with the contents of the cell. Cell division takes place through binary fission. In the eukaryotic cell, characteristic of all other organisms, the genetic material is divided into several chromosmes and separated from the rest of the cell by a membrane surrounding the nucleus. Transport may take place in both directions across this membrane. The cytoplasm contains numerous organelles: mitochondria, reticulum, Golgi apparatus. Division takes place through mitosis. With the eukaryotic cell, sexual reproduction and meiosis make their appearance

Cephalochordata. Lanceolate animals which are several centimetres long. The nerve cord is not differentiated into a brain in the frontal portion. The notochord extends along the complete length of the body. The pharynx is pierced by a large number of branchial slits. The outer body muscles show marked metamerism. This exclusively marine group comprises only a small number of species, such as the lancelet which lives buried in beach sands

Character. Sum of factors (molecular, physiological, structural, behavioural, etc.) making up an organism

Chordata. Characterized by, among other factors, the development of superimposed axial organs in the plan of bilaterial symmetry. From dorsal to ventral these are: the neural axis, the notochord, the digestive tract which in the head is widened into the pharynx from which, in turn, at least in the embryo, extend visceral pockets, and the heart in a ventral position below the pharynx

Chronocline. A feature evolving gradually with time along a certain gradient. The term chronomorphocline is frequently used to describe the gradual evolution of a morphological feature

Cladogenesis. Divergent evolution starting when a lineage splits into two descendant species. Cladogenesis was included in the synthetic theory as an evolutionary mode. In the synthesis introduced here it is considered as the consequence of the various phenomena leading to the formation of a species

Cline. Continuous and gradual morphological variation corresponding to a geographical environmental gradient (temperature, insolation, etc.)

Cnidaria. Mostly marine animals which are either free-floating (jellyfish) or sessile (sea anemones, corals). Some species such as *Hydra* also occur in freshwater. Two fundamental, very thin sheets enclose a gel-like substance, the mesogloea, which is occupied to varying degrees by cells. The "endoderm" surround the gastrovascular cavity. The jellyfish possess rather rudimentary sensory organs. The group derives its name from urticarian cells, the so-called cdnidoblasts. In older classifications the Cnidaria were considered as a branch of the Coelenterata, which is now subdivided into Cnidaria and Ctenaria

Codon. Sequence of three nucleotides in messenger RNA, the bases of which define a specific amino acid

Coelom. In the coelomata, this is a cavity generally lined by the mesoderm and enclosing the body organs (digestive tract, kidneys, and reproductive organs). The pseudo-coelom of the nemtodes is of a dfferent origin. It is derived from the cavity of the segmented egg

Consanguinity. Cross between genetically similar parents. Consanguinity favours the formation of homozygotes and may lead to misformations and degenerations when certain alleles exerting an unfavourable influence are present at double the number

Contingency/necessity. The two terms are contradictory, but do not have a strict meaning unless used comparatively with certain data. A fact or proposition is necessary when only this is reconcilable with the available data, or its is the only solution possible for the given proposal. A future event is contingent when, after a certain point, it may or may not occur. Evolution itself is contingent in as much as the course it follows after a certain points is not a necessity and is not predictable.

Deme. Local population

Demographic strategies. The combination of factors controlling the quantitative evolution of populations, like rate of fertility, mortality, age of sexual maturity, longevity, etc. Two types of strategy may be distinguished. *r* strategy: Acts as a factor of occupation of environments subjected to highly variable conditions. Selection here favours a high rate of reproduction, early reproduction, short life span, and small size of individuals. Mortality is high. An increase in productivity takes place at the expense of an immense "waste" of living matter.*K* strategy: Practised by organisms living under stable conditions where selection favours fertility and mortality to a different degree from that of the former strategy; a longer life span is favoured as well as slower rates of reproduction and growth. This strategy is used by stationary populations stabilized where selection favours efficiency

Diffusion. Physicochemical phenomenon of the spontaneous movement of molecules, without loss of energy, from a zone of high concentration to a zone of lower concentration. Diffusion may operate freely within a gas or liquid, or across a "wall" (see osmotic pressure)

Dimorphism. Variation within a population leading to the appearance of two morphologically distinct types (morphs). The classic example is the sexual dimorphism developed between males and females

Diploblastic/triploblastic. In the early stages of development the embryonic cells forms an external (ectoblast) and internal (endoblast) layer in diploblastic animals; the triploblastica have a third, intermediate layer, the mesoderm. The various organs form from these layers

Echinoderms. This phylum exhibits highly characteristic features including pentamerous radial symmetriy, and formation of a skeleton from superficial platelets arising from the mesoderm. The skeletal elements are either separated, as in the starfish, or combined in a shell, as in the sea-urchin. Exclusively marine, poorly mobile or sessile: sea-lilies (crinoids), starfish, sea-urchins, holothurians (sea cucumber, like the trepang of Chinese cuisine), brittle stars (ophiuroids). The echinoids are subdivided into the regular sea-urchins which have radial symmetry around a vertical axis, and the irregular ones which have bilateral symmetry

Ecophenotypic variations (Phenotypic Plasticity). These are repetitive and reversible morphological variations linked to fluctuations in environmental factors (e.g. European snails like *Pupilla*; Chap. 9)

Electrophoresis. Analytical technique based on the ability of a protein to migrate in an electric field to one of the two poles at a velocity controlled by its electric charge and molecular weight. With the aid of this method, it is possible to study the different forms of a certain protein in a sample derived from various individuals of a population

Epigenesis. During the 18th century two theories existed to explain the formation of organisms: preformation, where all structures are already present in the egg or spermatozoid, and only expand and grow during development; and epigenesis, where the organic complexity is established progressively during development. The concept of preformation has since been abandoned. The term epigenesis can have different meanings and the one adopted here (Chap. 4) is not the only one accepted today

Equilibrium (punctuated equilibrium). Evolutionary model introduced by Eldredge and Gould (1972), based on speciation according to the allopatric model of

E. Mayr. It asserts that species, once formed, remain stable and in equilibrium during their entire existence over millions of years. This equilibrium is disturbed only by a new phase of speciation. When the descendant species regains contact with the original form (parent species), it eventually replaces the parent species due to its higher competitiveness. This model explains the abrupt replacement of one species by another, which is well-documented in the palaeontological record. However, it is not exclusive as phyletic gradualism also frequently occurs in evolutionary lineages

Punctuated equilibrium/disequilibrium. Evolutionary model proposed by Chaline and Brunet-Lectome (1990) to complete the punctuated equilibrium model of Eldredge and Gould by introducing phenotypic plasticity and phyletic gradualism together with morphological stasis as other components of evolutionary modes (see Chapt. 9 and 11)

Foraminifera. See Protozoans

Fossil. A term originally describing only a curious mineral formation. The present definition was given by Lamarck: organic remains, animal or plant, contained in sedimentary formations. Pseudo-fossils are traces of organic activity, such as burrows, stromatolites (laminated calcareous structures formed by cyanobacteria), etc.

Founder Principle. Introduced by E. Mayr. A group of individuals, of varying numbers (it may even consists of a single fertilized female) becomes detached from a larger population and starts to colonize a new environment. The group may form a new population which, in turn, could lead to a new species (allopatric speciation)

Gametes. Reproductive cells, male or female, formed from parent cells, the gonia. The reproductive cells constitute the germ cells, as opposed to the somatic cells which make up the remainder of the organism

Genome. Genetic content of chromosomes, i.e. the array of genetic programmes contained in the DNA

Gradualism. Concept according to which an evolutionary change of any extent may take place only through small changes leading in the same direction. This is one of the underlying concepts of the synthetic theory. In contrast to this exclusively gradual evolution, authors such as Goldschmidt and Schindewolf have stressed the possibility of discontinuous evolution by morphological jumps (saltations), i.e. the passage from one body plan to another without intervening stages. According to Goldschmidt, such jumps would be triggered by "systemic mutations" necessitating a considerable reorganization of the genetic programme. They lead to the appearance of "hopeful monsters" at the start of new lineages. Although there is no proof for Goldschmidt's thesis, the possibility of discontinuous evolution under certain circumstances can not be categorically excluded (Chapt. 6).

Hierarchical Levels of Organization. Living matter is arranged in a series of organizational levels (the "keyboard of living matter"). There is a hierarchy of increasing complexity from the genetic level through various states of development and morphology to the ecology and ethology of a species. Divergence between two populations may appear at any of these levels, and eventually lead to reproductive isolation. The hierarchical arrangement of these levels leads to an increase in complexity from one to the next, with the appearance of new properties which cannot be deduced from the structures at lower levels

HLA System. Groups of antigens called "histocompatible" and which are fixed to the surface of cells, representing their "visiting cards". When introduced into another organism, these antigens induce the formation of antibodies which destroy the intruders. The HLA system is one of the factors responsible for the rejection of organ grafts. It makes the individual immunologically unique

Hybridization of DNA. A technique facilitating an estimate of the degree of similarity between the DNA nucleotide sequences of different organisms. It entails the recombination of DNA fragments from a single strand of the double helix of different species. The completely or partly corresponding sequences regroup in hybrid double chains which may be recognized

Hypermorphosis. The delayed onset of sexual maturity allows growth to continue for a longer time. As a consequence, the adult will be of a larger size with a "hyper adult" morphology

Inoceramus. A Mesozoic bivalve mollusc

Isolate. A group of individuals of a particular species which becomes completely isolated from the remainder of the species without any further genetic interchange (suppression of genetic flux). For animals, the causes of isolation are generally of a geographical nature. A little-known example is the crocodiles of the Tassili N'Ajjers marshes of southern Algeria which became isolated from the rest of the African crocodile population by the formation of the Sahara. The last of these crocodiles disappeared only very recently. For man, isolation may be caused by social or religious factors, as in the case of the Amish sect in eastern Pennsylvania (USA)

Karotype. Chromosomal complement of species

Macroevolution. Evolutionary developments beyond the individual species, leading to genus, family, etc. The evolutionary trends and radiations inherent to a given body plan

Megaevolution. Appearence of body plans (phyla)

Metabolism. Sequence of molecular transformations taking place within an organism and ensuring its continued existence. Anabolism refers to the formation of larger molecules from smaller ones, while catabolism describes the reverse process, such as occurs during respiration

Meiosis. During formation of the sex cells (gametes), the two division of meiosis take place. Starting from a basic cell with 2n chromosomes, the first (reductional) division leads to two daughter cells which each contain half the number of chromosomes (haploid number, n). During the second (equational) division, each cell produces two further cells with n chromosomes. For man, the diploid number is $2n = 46$ and the gametes contain $n=23$ chromsomes

Metamerism. A constructional principle which entails repetitive patterns of elements along the entire length of an organism. A truly repetitive construction is not possible in nature and metamerism has become modivied to some degree. Thus, in certain annelids which exhibit the highest degree of metamerism, only the anterior and posterior segments of the body differ from the metameric majority of the body. As already clearly evident on the outside, the individual metameric segments are separated by a diaphragm. Structures like the ganglia of the nervous system, the coelom, and the excretory organs are well metamerized, whereas the digestive tract is not. In the earthworm, the reproductive organs are present in only a few of the metamerized sections. In the crustaceans, several anterior segments combine to form the cephalothorax. In the vertebrate embryo, the lateral body muscles, the nerve cord, the vertebral arteries, and the position of the excretory organs are repeated to a varying degree

Microevolution. Processes and mechanisms explaining the variability of species, their formation, and their evolutionary patterns, but limited to intraspecific changes

Mitosis. The start of a complex multiphased process leading to duplication of the chromosomes and in which each daughter cell ends up with the same number of chromosomes as the mother cell (diploid number, 2n).

Molluscs. Subdivided into several classes of which the following are of special interest here: gastropods – enclosed in a simple shell; the prosobranchs are the most primitive *(Patella, Littorina),* and the pulmonates (snails) breathe through a form of lung. Lamellibranchs or bivalves – with two parts to the shell, a left and a right valve. Cephalopods – the foot is subdivided into arms; they are the most complex of all molluscs and are exclusively marine (octopus, squid, nautilus, ammonites)

Neoteny. Reduction in the rate of morphological development affecting either the entire organism or particular structures, and leading to an adult which is identical in size to its ancestor but which conserves some ancestral juvenile characters

Nutrition. Autotrophic nutrition is found in chlorophyll-bearings plants which synthesize their organic components from mineral compounds with the aid of solar energy. Animals are heterotrophic, degrading organic compounds derived from plants or other animals in order to grow. Osmiotroph absorb organic compounds from the environment accross their body surface. This probably represents the type of nutrition of the first organisms prior to the appearance of chlorophyll.

Ontogenetic Trajectory or Itinerary of Development. The succession of the different stages of development of a structure (or a complete organism) may be represented in a simplified fashion by the curve in a coordinate system with three axes: one representing age (time), one size, and the third shape (ratio of two dimensions such as length and height). This method of representation was introduced by Alberch et al. (1979)

Ontogeny. Arbitrarily stopped when an organism is able to enter its environment. It should actually also entail all later transformations and thus cover the complete life cycle of an organism. In a more restricted sense, organogenesis covers the period during which the organs are formed after gastrulation. Here it is used in a sense equivalent to embryonic development, as this is more familiar to the general reader

Operon. Concept introduced by F. Jacob and J. Monod. In bateria the genes do not act independently but are associated in autonomous functional units (operons) made up of structural genes and their regulatory genes.

Osmotic Pressure. Best illustrated by an example such as that of a cylinder closed by a piston and subdivided by a membrane into compatements A and B of equal volume. A contains pure water, and B a solution of 180 g glucose in 1000 g water. The so-called semipermeable membrane allows water to pass across it much more rapidly than the sugar. Water diffuses from A to B which consequently increases in volume. To stop the flux of water we must exert on the piston closing off compartment B, a pressure of 22.4 kg/cm^2 this corresponds to the osmotic pressure of the sugar solution against pure water. However, the sugar also gradually diffuses across the membrane in the opposite direction, and the concentration in the two compartments eventually becomes equal. In vertebrates, the osmotic pressure of the blood, caused essentially by the content of sodium chloride would be constantly modified, either by exchange across the skin in the aquatic environment (fish) or by metabolic activities (fish and terrestrial animals) if the osmoregulatory mechanisms were not continuously active

Paedomorphosis. Reduction in the degree of development over geological time, leading to adult descendants conserving the juvenile morphology of their ancestors. Paedomorphosis may result from neoteny, progenesis, or post-displacement

Parapatric. Two species are parapatric when their geographical areas of distribution are contiguous

Pelagic. Living in the open sea and far above the seabed. The bulk of pelagic organisms makes up the plankton

Peramorphosis. Increase in the degree of development over geological time, leading to a new "hyperadult" morphology resulting either from acceleration, hypermorphosis, or pre-displacement

Phenotype. Group of characteristics of an organism which are the expression of part of its genetic programme, controlled by the mechanisms of development and subjected to the constraints of selection. Here it is used synonymously with the term organism

Photosynthesis. Process occuring in certain bacteria and plants containing chlorophyll, a pigment capable of capturing from solar light the energy necessary for organic synthesis using mineral (green plants and certain bacteria) or organic compounds (certain bacteria). Aerobic photosynthesis utilizes carbon dioxide and water to make sugars; oxygen is liberated in this process. Anaerobic photosynthesis does not produce oxygen. It is employed, for example, by sulphur algae containing bacteriochlorophyll; these use carbon dioxide and hydrogen sulphide, and sulphur is produced during this process

Phylogeny. Evolutionary history of lineages and groups of organisms

Plankton. Organisms in sea or fresh water which float near the surface and are more or less passively carried by the currents. Includes unicellular animals and plants, cnidaria, larvae of annelids, molluscs, crustaceans, etc.

Platyhelminthes. The former "flat worms". This group includes the freely mobile freshwater planaria and the parasitic fluke in the liver of sheep

Polypeptides. See proteins

Post-displacement. Delay in the onset of growth of an organic compared with the overall growth of the body leading to an adult with a juvenile stage of the respective organ

pre-displacement. Precocious onset of growth of an organ compared with the overall growth of the organism, leading to an individual with an advanced morphology of this organ

Principles of Metazoan Classification. Species: The basic unit is the biological species, defined by the ability to interbreed. "Species is a group of populations which may interbreed with each other and are reproductively isolated from (non-interbreeding with) neighbouring groups. A species occupies a particular ecological niche." (E. Mayr)

The species problem: the above wording defines a category without referring to any organism in particular. In considering an example such as the wolf, we must describe a taxon "species" as well (in this case, *Canis lupus*).

The definition of the biological species is based only on reproduction as a criterion. It ignores morphology as well as distribution in time and space. It defines an extant species. When morphology is also taken into account it loses it general character, as in the living world (and certainly amongst fossils) we know of examples which may not be distinguished morphologically, and which would, at first sight, be grouped together in the same species although they do not interbreed. They belong to different species, the so-calleds sibling species of which examples are known from a number of different groups.

Let us now introduce the parameter "space: Throughout its area of occurrence, a species may exhibit characteristic regional differences. Such species are called polytypic, as opposed to monotypic. The black tit *(Parus major)* occurs unchanged throughout Europe and northern Asia, from Brittany to Kamchatka. Towards Iran, a particular population or subspecies *(Parus major cinereus)* develops from the basic species. Towards India and Indochina, there is another subspecies *(Parus major minor)*.

The next parameter is "time": The problem becomes complicated as the criterion of interbreeding loses its significance.

Let us consider an annual species, the common housefly *(Musca domestica):* when in the autumn of 1985 the flies disappeared, several specimens went into hibernation and, on re-awakening the following spring, formed the basis of the 1986 generation. Thus, the flies of 1985 and 1986 are connected by interbreeding and belong to the same biological species. This criterion, however, does not apply to distant generations of flies, for example those of 1985 and 1989.

Based on the evaluation of illustrations and collections, we may say that all these flies resemble each other and that the species had remained unchanged since the 18th century (or even earlier) up to 1989 (and possibly also thereafter). In this case, morphological similarity is the important criterion.

One or two centuries are negligible compared with the dimensions of geological time counted in hundreds of thousands or even millions of years. We must admit that in the past, the appearance of a species has remained the same as we observe it now and that a species may remain stable over considerable periods of time. How then may we trace the limits of a species through time? How do we delineate its particular life span? This is the highly complex problem of the palaeontological species. Let us assume that from the same horizon of a sedimentary deposit we collect various ammonites which resemble each other so much that we may assign them to the same species. We are, however, certain that they were not exactly contemporary and that between the existence of two samples thousands of years may have elapsed. One or two metres higher up we find what appears to us to be the same shell and thus the same species. In order to establish that this is really the case and that we are dealing with a series of populations (biological species), interconnected by successive steps of interbreeding, we have to take recourse in the statistical treatment of a large number of mainly measurable morphological features. The statistical tests will then allow us to evaluate whether or not we are dealing with only one species.

In this book we have developed a spatial-temporal species concept which integrates the time dimension with the biological species definition: "A species represents the spatio-temporal continuum between natural populations which at any point of the continuum may interbreed with each other, and which are reproductively isolated from any other analogous group". Under this concept, a species will last from the attainment of reproductive isolation (speciation in the biological sense) through to its extinction. The longevity of species may vary, but all are characterized by the phenomena of stasis, ecophenotypic variation (or phenotypic plasticity), and phyletic gradualism. Palaeontological species are exclusively morphospecies, frequently referred to as palaeospecies, corresponding to "instants" subdividing the continuum of a chronospecies into successive "evolutionary steps". Under good conditions of fossilization, morphospecies may approach true biological species, although recognition of sibling species is not possible by statistical methods.

Morphology is not used exclusively by palaeontologists. A zoologist will also use it in the case of a collection of specimens where it is difficult to recognize interbreeding "in action".

Classification beyond the species: Various species may be grouped together as a genus which, however, we cannot define as a category by general characteristic features. Describing a taxon is the task of a specialist of a certain group. Under the genus *Canis* (here a taxon), we may group the wolf *(Canis lupus)* and the jackal *(Canis aureus)*. However, the extension of a taxon is quite arbitrary and, furthermore, is handled differently by different authors. There are also other animals which, in a more remote way, resemlbe the wolf and jackal; consider the desert fox *(Fennecus zerda)* and the wild hunting dog *(Lycaon pictus)*. The genera *Canis, Fennecus,* and certain others constitute the family (taxon) Canidae.

The Canidae, in turn, may be grouped together with the Felidae (lion, tiger, cat, etc.) in the order (taxon) Carnivora. By successive regrouping according to a decreasing degree of similarity, we arrive at the class (Mammalia), the phylum (Vertebrata), and eventually the kingdom. A number of less important intermediate categories have been omitted here.

Starting from features characterizing the wolf, we have progressed to increasingly more general features. Such is the basic principle underlying all classification or taxonomy.

Polypeptides. Short chains of amino acids connected by peptide bonds between the acid group of one amino acid and the amine group of the adjacent one. The arrangement of amino acids in certain sequences makes polypeptides code-bearing molecules

Progenesis. Change in development of an ancestral organism by the early appearance of sexual maturity which prematurely stops growth and thereby results in a small adult with the juvenile morphology of its ancestor

Protein. Amino acids are simple organic molecules combining the function of an acid with that of an amine. There are 20 different amino acids which make up all natural proteins. Proteins are then formed by the combination of several polypeptide chains (and eventually also of other molecules). They may become structural proteins, the building blocks of cells, or enzymes, i.e. catalysts operating at body temperature

Protostomians/Deuterostomians. In the former, the blastopore develops into the mouth of the adult, or into the mouth and the anus, or the two may open up again as secondary features on emplacement of the blastopore. Segmentation is basically of the spiral type, with exceptions like the cephalopods. The destiny of the cells (blastomeres) is determined rather early (mosaic development). In the nerve cord, the cells become concentrated into ganglia. The nerve cord is on the ventral side of the digestive tract, passing above it only in the anterior portion of the body. In the deuterostomians, the anus forms from the blastopore or opens up as a secondary feature on the emplacement of the latter. The mouth represents a neo-formation. The destiny of the blastomeres is not strictly determined until a rather late stage (development by regulation). Segmentation is of the radial type. The nerve systems forms a network below the epidermis in echinoderms, whereas in the Chordata it forms a tube on the dorsal side of the notochord and digestive tract

Protozoa. Unicellular organisms in marine and fresh water. The phylum is subdivided into several groups such as the radiolarians with a siliceous skeleton that is frequently finely sculptured, and the exclusively marine foraminifera which possess an outer and inner (calcareous) shell, sometimes subdivided into chambers

Radiolaria. See Protozoa

Reductionism. Concept according to which process or construction of any complexity may be explained only by the properties of its constituent elements. It thus totally ignores anything that may intervene between the component level and the realization of a process or organism. Attributing all problems of evolutionary change only to genetic programming represents a reductionist position which does not takes into account processing of the genetic information during ontogeny or the emergence of new properties as the result of the constructional or integrational levels successively achieved during development

Replication. Auto-reproduction of a DNA molecule into two daughter molecules (two double helix strands)

Reptiles. Class made up of several recent and fossil groups, like tortoises, lizards, snakes, crocodiles, sphenodontia (e.g. *Sphenodon,* a kind of lizard in New Zealand), dinosaurs, ichthyosaurs, pterosaurs or flying reptiles, etc. It is a rather heterogeneous assembly without any common characteristic defining the class. Cladists stress the artificial basis of this grouping and divide it into several classes. Here the well-established term "reptiles" is retained although the systematic significance has been lost. To indicate this quotation marks are used. The fish, a similarly artificial "class", should be considered in the same way

Reproductive Isolation. The ultimate consequence of divergence between two populations, leading to the establishment of two new species. Isolation may set in at any organizational level (genetic, biochemical, ecological, ethological, etc.). It precludes interbreeding when the two isolated populations come into contact again

Respiration. See aerobic/anaerobic

Rhesus Factor. Group of blood antigens. When a foetus inherits rhesus antigens from its father, the contact of the foetal blood with that of the mother leads to the formation of antibodies in the latter and these destroy the red blood cells of the foetus

Saltationism. Concept according to which evolution may take place in jumps without intermediate forms

Sequencing of proteins. Determination of the sequence of amino acids in a protein

Speciation. Formation of a new species

Species. See principles of metazoan classification

Spongiaria. Sponges

Sympatric. Two species are sympatric when their areas of distribution overlap, at least partly (see also Allopatric and Parapatric)

Systematics. In a very general sense, systematics deals with the diversity of the organic world. It makes use of various methods (see also Principles of Metazoan Classification).

Definition of categories ordered in a hierarchical system. Introduced by C. Linné, the system uses, in ascending order: species, genus, family, order, class, and phylum, together with numerous intermediate categories.

Classification sorts organisms into groups according to their relative similarities. The nomenclature fulfils, as it were, the function of a magistrate's court by giving names to organisms according to rules laid down in a code.

A category is *defined* by certain general properties without considering any organism in particular. Thus, the category "species" is defined by the criterion "interbreeding" (Chap. 7), whereas a phylum is defined by the presence of a certain type of construction or body plan.

A taxon is a group of organisms characterized sufficiently to be given a name and assigned to a category. A taxon is *described* by particular characteristics. Thus, in the category "species" we may include and describe the taxon "grey rat" *(Rattus norvegicus)* on the basis of certain characteristics which distinguish it from the taxon "black rat" *(Rattus rattus).* At the level of the genus, dog, wolf, and fox belong to the taxon "*Canis*". In the category "class", the taxon vertebrates is described by common features like an internal skeleton, brain, spinal cord, closed blood circulatory system, etc.

Transposon. Mobile genetic sequence which, as a jumping gene, may spontaneously move from one position to another on the same chromosome or to a different one. Discovered in maize (corn) by B. McClintock (1951), transposons are now known from a number of prokaryotes and eukaryotes. They may play a role in the triggering of evolutionary changes like speciation (?)

Tunicates or Urochordates (Ascidia). Exclusively marine group, the adult forms of which are either solitary or occur in free-floating (pelagic) or sessile colonies. The larvae possess axial organs similar to those of the Chordata. During metamorphosis, certain of these organs degenerate and the adult then forms a sac with an enormous pharynx (for respiration and the gathering of nutrients). They also have a short digestive tract, contractible pocket (heart), genital organs, and an excretory organ. The chordate construction principle is no longer recognizable

Vertebrates/Invertebrates. These two terms are not of similar significance. Vertebrates are a phylum (taxon) which is well defined by a general body plan. In contrast to this, there are no common positive characteristics describing such different invertebrates as a sponge, mollusc, or insect. The term invertebrate, introduced by Lamarck for animals without vertebrae, covers various phyla of rather different body plans

Viscero-Somatic Equilibrium. The body functions of an animal may be subdivided into two basic groups:

The visceral system entails processes, like the gathering and processing of nutrients and a supply of energy for the functioning of the organism; such processes include digestion, respiration, circulation, and excretion.

With the somatic system, the organism perceives and responds to its environment. This involves the sensorial organs, and nervous and muscle systems.

The reproductive system constitutes a separate category.

In an ascidian, a sessile animal, the visceral systems predominate: enormous pharynx gathering nutrients and respiratory compounds, short digestive tract. The somatic system is reduced to a small ganglion.

In the evolution of vertebrates, the development of mobility is, amongst others, an important factor for the viscero-somatic equilibrium. The vital visceral system remains intact, but the somatic system grows in importance with the development of sensorial organs, the head and brain, and the locomotive muscles

Appendices

Chapter 3. Genetic Material and Programmes

Appendix 3.1. Some Genetic Information

The genetic material is contained in the chromosomes of the cell nucleus. From a functional point of view, the material is subdivided into particles, the genes. Their activity controls the formation of proteins which are either enzymes, catalysts of the biochemical activities of the cell and its metabolic cycles, or structural proteins, the building materials of cells.

Each gene is present in the cell in two copies, one of which, on fertilization, is introduced by the spermatozoid while the other is provided by the egg cell. The same gene may occur in different forms (alleles) in the various individuals within a species. These alleles do not possess exactly the same activity. In *Drosophila,* for example, the gene called w by geneticists controls the colour of the eyes, which may be brown-red, vermillion, apricot coloured, etc. This proves that gene w exists as a number of different alleles.

By the term locus we refer to a defined position on a chromosome occupied by the various alleles of a gene in different individuals.

The totality of the genes of an organism constitutes its genome (or genotype), the final organism being the phenotype.

The genetics of populations utilize the concept of a gene pool, containing all the genes of a population withouth differentiating between them with regard to their exact parental origin or their generalogy.

Appendix 3.2. Chromosomal Mutations

Changes in the number of genes are caused either by removal (deletions) or multiplication (duplications). The changes in position (Fig. 5.5) are: inversions, during which a block is rotated by 180; or translocations, which result in a change of the gene position on the same chromosome. Blocks may be exchanged between the same pair or between different parts of chromosomes. The change in position of a gene or a block of genes modifies their genetic surroundings and may thereby influence their activity (position effect).

Changes in the number of chromosomes: The nucleus of a eukaryotic cell contains n pairs of chromosomes, i.e. 2n chromosomes (diploid number), whereas the sex cells contain only n chromosomes (haploid number). Two chromosomes may coalesce (fusion) or one chromosome may split into two (fission). One or more chromosomes may be missing entirely (anaeuploid) or be present more than once. In mongoloid humans chromosome 21 is present in triplicate. In the polyploid case, the cell contains 3n (triploid) or 4n (tetraploid) etc. chromosomes instead of the usual 2n.

Appendix 3.3. Rates of Mutation and Evolutionary Capacities

At what frequency may a certain locus mutate?

To establish this frequency is rather difficult as it requires a large number of observations. The rates are always low, in the order of one mutation for every n million gametes in man, as well as in *Drosophila,* or in maize (corn).

To what extent will the rates of mutation influence the capacity or rate of evolution? Geneticists studying populations tend to agree that the effect of the rate of mutation on the frequency of alleles is so small that it will only play a minor role in affecting the direction of evolution.

Due to the accumulation of mutations during their history the gene pools of populations will always contain a sufficient degree of genetic variability to respond to multiple selective influences. Thus, no new mutations are necessary when a population has to face an environmental challenge.

It appears that at the genetic level, evolutionary mechanisms rely more on changes in the regulatory genes than in the structural genes.

To evaluate evolutionary mechanisms we have to distinguish between the molecular and organic levels. For evolution at the molecular level, i.e. in nucleotides and codons, mutations (s.l.) play a decisive role, as they can create novelties.

At the organic levels, selection and genetic drift will intervene, although neither may create novelties in the true sense of the word "create". Molecular variability, on the otherhand, will not take effect unless translated into integrated structures in an organism which is subjected to selection in a particular environment.

Appendix 3.4. Frequency of Alleles

In a population with N individuals, a pair of alleles a1–a2 may form three genotypes: a2 a2; a1 a2; a1 a1, the frequencies of which are X, Y and Z respectively, or X+Y+Z=N. The total number of alleles is 2n (two per individual). The frequencies (p) for a1 and (q) for a2 are:

$$p = (2X+Y):2N,\ q=(Y+2Z):2N,\ \text{or}\ p+q=(2X+2Y+2Z):2N=1.$$

To explain the changes in allele frequency in the gene pools of successive generations, one also has to take into account the selective value of a certain allele.

Appendix 3.5. The Neutralist Theory of Evolution

This rejects the selectionist foundations of Darwinism. Like evolutionary genetics, the neutralist theory analyses evolutionary changes on the basis of the gene, isolated from the bulk of the genetic material. When, however, the causes for the frequency of changes in the genes within populations and over generations are considered, the two approaches diverge completely.

The selectionist theory is a deterministic one: The appearance of the various forms of the same gene (alleles a1 and a2) is controlled by their relative selective values. When the selective value of a1 is slightly higher than that of a2, the former will replace the latter over a number of generations.

According to neutralism, the frequency of the gene changes within a population and over the course of a number of generations results from the random phenomenon of genetic drift [11]. At the molecular level, at least, selection does not play a role. This led Thomas H. King and Jack L. Jukes [6] to talk critically of a non-Darwinian theory of evolution. The calculations, as well as the theoretical considerations, which would justify such positions cannot be described here, and we will limit ourselves to the basic proposals: Considering the observed or calculated rates of mutations, it is concluded that if each mutation had an effect on the organism concerned and was thus subjected to selection, the organism would not be able to sustain this quasi-permanent pressure because of the adaptive repercussions. Most, if not all, must therefore be neutral and not the target of selection. Variations in allele frequency thus result only from division and accidental redistribution during the formation of the spermatozoids and egg cells, as well as during their encounter at fertilization.

According to the proponents of the neutralist theory, this would also explain a number of observations on the distribution of genes in populations, upon which the selectionist theory is based. We would point out, along with Kimura, that neutralism only applies at the molecular level and that at the higher level of the organisms themselves, selection will indeed operate.

As neutrality is only a theoretical conclusion, it should be able to directly prove or disprove it by facts. Thus, selectionist geneticists look for, and manage to find here and there, some structural and functional repercussions of mutations which they consider to be neutral. The alternatives boil down to showing that neutralist and selectionist theories are mutually exclusive, or to admitting that they are compatible with each other and are, at the same time, complementary.

However, as Jacques Ruffié [19] has explained, neutralists and selectionists do not speak the same language.

Neutralists consider the individual gene and its mutations, whereas selectionists with the same justification look at groupings within gene populations, these groupings constituting the genetic material.

The neutralist investigations agree with those of the founders of evolutionary genetics in understanding the organism as a collection of genes which may be isolated, and the behaviour of which may be studied like that of independent objects. They thus adhere to a somewhat "atomistic" concept of an organism.

In the living world, natural selection acts *directly* on an organism and in this way also *indirectly* on its genes. Dobzhansky has pointed out repeatedly that all genes of a certain genetic material form a functional assembly in which they have a different selective value despite their *simultaneous* presence in the same genetic ancestor. They represent a *coadapted systems.*

The variation in detail of the molecular level, made up of the genes themselves, will not in the majority of cases find expression in the organism concerned. We know, for instance, that in a given species there are no identical individuals (with the exception of identical twins) that possess exactly the same genetic material.

However, the individuals still resemble each other in their general features of the same selective value and are able to interbreed (Chap. 7).

This shows that the variations at the molecular level are largely evened out or "soaked up" by the functioning of the genome itself and by the mechanisms of its realization during the development of the embryo (Chap. 4).

The selective pressure forcing an organism to adapt to its environment acts on the organism as the material expression of combinations of genes belonging to one or more coadapted systems, of combinations which cannot be selectively neutral. The birth of a new species then corresponds more to the establishment of a new genetic arrangement than to the action of new mutations.

In conclusion we may state that the neutralist and selectionist theories do not contradict each other. They should be considered simultaneously at two "successive" levels, the molecular and the organic; this makes the contradiction more supposed than real.

Chapter 4. From Egg to Adult: Development

Appendix 4.1. The Development of *Drosophila*

Without entering into detailed analysis we may conclude:

- Genetic determinism unleashes a cascade of impulses from the activating genes onto the selective (or commutative) genes which eventually control the structural genes of the two complexes.
- We must consider the complexes as consisting of genes and groups of cells which control what we refer to as the regulatory systems of development (termed "compartments" by Garcia Bellido [15]).

With the advance of organogenesis, the regulatory systems progressively split into subsystems which become more restricted and more precisely directed towards a certain task. The system regulating the second body segment initially covers the entire extent of the future segment. Then the antennapedia gene complex starts to determine certain subsystems for imaginal discs, groups of larval cells which, during metamorphosis to the adult form, will result in a wing, leg, antenna, etc. Within such a disc, dorsal and ventral portions then form, and in each of these anterior and posterior sections. Eventually we find systems expressing themselves in differents parts of, for example, a wing during metamorphosis.

In this succession of restrictions we encounter genes which can be called commutating or selective genes, as their action puts an alternative before the future differentiation of a group of cells: will it develop into a dorsal or ventral portion, into an anterior or posterior section, etc. [16].

Appendix 4.2. Heterochrony in Development

Axolotls fed with thyroid glands from, for example, calves, may metamorphose into terrestrial ambystomes, provided that they have not yet laid eggs: the attainment of sexuality inhibits metamorphosis (Fig. 4.9).

We now subdivide the life cycle of the ambystome more or less arbitrarily into a sequence of events S1, S2 ... Sn, ... M, ... G. The period S1 to Sn represents the life span of the larva. After the event Sn, metamorphosis (M) sets in, followed by the maturity of the sexual organs or gonads (G). The axolotl thus follows the initial steps of the sequence, but G directly follows Sn, as metamorphosis no longer occurs. This phenomenon is called neoteny or reproduction in the larval stage, and results from the duplicate presence in the homozygote stage of a recessive gene P, which suppresses the production of the thyroid hormone. The action of this gene is controlled by the temperature of the environment and is counteracted by a supply of thyroxine.

The intervention of such time-controlling genes has also been recognized in the development of the nematode *Caenorhabditis,* a sort of small "worm" which has only 1000 cells in the adult stage.

Appendix 4.3. Relative Growth

The relationship of relative growth may be expressed as a mathematical formula:

$$y = bx^{\alpha}$$

Where b is a constant derived from actual measurements and α is the growth coefficient of y compared with x.

Transposing the formula into logarithms, we obtain

$$\log y = \log b + \alpha \log x,$$

defining a regression line with the slope α intersecting the y axis at log b. The coordinate system starts at $x = 1.0$, corresponding to $\log x = 0$.

If $\alpha = 1.0$, we obtain $\log y = \log b + \log x$, and y and x grow at the same rate. The two dimensions are multiplied with the same coefficient (isometric growth) and the shape is preserved, as in similar triangles.

If α differs from 1.0, the two dimensions are multiplied with different coefficients and growth will be differential or allometric. The shape will not be preserved.

If α is below 1,0 y will grow at as lower rate than x. This is the classic case in human development where the size of the head, compared with that of the rest of the body, is much larger in an embryo or child than in an adult. If α is above 1.0, allometry is positive and y becomes successively larger than x. The study of the head during the development of the domestic horse and during the phylogeny of the lineage over the Tertiary supports these theoretical considerations [32]: Let us take y as the length of the face and x as the length of the brain case (Fig. 4.12). The ontogenetic growth may be represented by a second-order equation describing the branch of a parabola: $\log y = -0.31 \log x^2 + 2.42 \log x - 1.51$.

This curve shows that the relative growth coefficient (α) is not constant, but decreases during development from 1.58 in the initial stages to 1.0 (isometry) after birth and during juvenile growth.

In the evolutionary lineage of horses (Fig. 4.13), from the small *Mesohippus* of the early Tertiary (Oligocene) to the forms around the Tertiary/Quaternary boundary which were like the present wild horse in size, the growth coefficient (1.0) was virtually equal in juveniles and adults, as in modern horses.

Chapter 7. The Formation of Species

Appendix 7.1. The sibling Drosophilas of Central and South America

The distinction between *Drosophila pseudo-obscura* and *D. persimilis* of Central and South America presents a classic example. Shortly after the species *D. pseudo-obscura* was described, it was recognized that it included two types of individuals, namely, A and B. Pure crosses between A and between B lead to fertile descendants, whereas crosses between A and B, where the two forms coexist, result in sterile males, in quasi-sterile females and a fragile progeny. The two differ in the nature of their enzyme system, in their chromosomes, and in chronological stages of development, as the females of A attain sexual maturity after 32–36 hours while the B females reach this stage only after 44–48 hours. Other differences concern the metabolic rhythms, with B being more active in the morning and A in the afternoon, and the shape of the sex organs and wings. Their nuptial displays are different as are the frequencies of the songs emitted by the males. As these appear to excite only the females of their own species, the possibility for crosses between A and B is virtually nil. The two species also differ in their ecological situation, with A preferring higher temperatures and B lower temperatures; consequently they encounter each other only in intermediate zones.

Chapter 8. The Historical Framework of Evolution

Appendix 8.1. The Primordial Atmosphere

The composition of the primordial atmosphere is still very much disputed. It could have been made up of carbon dioxide, water vapour, ammonia, hydrogen, and possibly also hydrogen sulphide, methane, etc. It was most certainly devoid of free oxygen which would have immediately destroyed by oxidation the newly formed molecules. The atmosphere was thus reducing and became oxidizing only somewhat later when the photosynthetic activity of chlorophyll-bearing unicellular organisms had liberated enough oxygen. At the same time an ozone layer started to develop at high altitudes; this absorbed the majority of ultraviolet radiation which is detrimental to all forms of life. The first organisms were protected from this radiation as they lived in water. Only when the ozone layer had become established as a virtual "shield", was life able to successfully venture onto land. That primordial molecules, like cyanide, formaldehyde, or acetylene, should have formed in this primitive atmosphere has lately been questioned. The primordial organic molecules may have been synthesized over billions of years in the solar nebula [6].

Chapter 9. The Species in the Course of Geological Time

Appendix 9.1. Other Examples of Gradual Evolution

Radiolaria: Drill hole E 14-8 in the Arctic Ocean intersected a continuous sequence over 2 million years during the upper Pliocene. It illustrates variations in size of the radiolarian *Pseudocubus vema* [14]. Unfortunately this evolution within the same lineage could not be observed in other boreholes in the area of distribution of this species. This would have been of prime importance, as size is a factor known to vary locally in unison with environmental factors like temperature, salinity, supply of nutrients, etc. Despite these gaps, the example shows that within this species there is a global increase in body size in a zigzag fashion. The increase is thus reversible and fluctuates around a general trend. The increase essentially takes place in three phases which were interpreted by S. J. Gould and N. Eldredge as three stages of stases separated by rapid changes. This step-like advance cannot be related to the model of punctuated equilibrium which implies the formation of species between phases of stasis, a situation certainly not applicable here.

In another example, Kellog [15] has shown that there is a gradual decrease in size of radiolarians of the genus *Eucyrtidium* in a borehole from the same zone in the Arctic Ocean (DSDP drill hole V 20–105). It also illustrates the formation of the species *matuyamai* from the species *calvertense*.

Ammonites: The ammonites constitute one of the principal groups from which numerous examples of phyletic gradualism have been described [16–22]. One of the better documented examples is the evolution of specific lipoceratids called "capricornes", a group of ammonites from the lower Liassic of northwestern Europe [19]. The complex history of this group consists of a phase of gradual evolution from *Aegoceras maculatum* to *Oistoceras figulinum;* this phase is observed in stratigraphic sections from different geographical areas. It is based mainly on the ornamentation of the shells.

The mammals of Wyoming: Based on detailed investigations in the Bighorn and Clark Fork Basins of Wyoming in the northwestern United States, P. Gingerich [23, 24] described a number of examples of gradual evolution, especially of the condylarthri, a group of primitive early Tertiary mammals. The changes in length and width of the teeth are indeed gradual in *Hyapsodus* as well as in *Haplomylus* or *Pelycodus*. Although the examples of Gingerich are properly documented from the well-preserved stratigraphic succession, they are lacking the spatial component. One would like to know about variations in the same parameters in other sedimentary basins, in order to find out whether one is really dealing with gradual evolution or only with regional intraspecific variations.

The musk rats of North America: This lineage, leading from the archaic ondatra *Pliopotamys minor* to the Recent *Ondatra zibethicus* [28] exhibits an increase in the dental height, i.e. case of hypsodontism, such as occurs in the mole rats. However, there is a greater increase in size and a more pronounced modifications of the tooth surface. In contrast to this, neither the disappearance of the dental roots nor the continuous growth are observed. This underlines the high degree of evolutionary independence of individual lineages.

Chapter 10. From Species to Body Plans

Appendix 10.1. The Reptiles Which Became Mammals

Reference is made here also to Fig. 10.1. Palaeontologists have described a succession of adult forms which led to the discovery of what is now referred to as sprouting evolution.

Among the numerous families of mammals-like reptiles which managed to achieve one or other of the mammalian features, only one was really successful.

These adult forms are the expression of embryonal developments which became modified with time. The recognition of the complete evolutionary lineage thus poses the question "How"?. How does development change? How does this tendency, starting from certain minor features in the skull of all early mammal-like reptikes, persevere and eventually become amplified in the radical "breakthrough" marking the end of the process some 100 million years later?

Fossils do not provide sufficient data to answer this question, but present animals may to some extent alleviate this difficulty.

In the cartilaginous skull of very young lizard embryos, the lower jaw is a small rod, widened at its posterior end, which touches a detached piece of the skull. The rod then degenerates, except for its posterior portion which becomes ossified into the articular bone. The detached piece becomes the quadrate and the mandibular articulation of the reptiles (articulo-quadrate) is established. The dentary bone grows backward against the articular bone, with which it fuses to form the skeleton of the lower jaw.

A very similar arrangement is also observed in the embryos of kangaroos and hares. The posterior portion of the cartilaginous rudiment becomes the malleus (hammer) which loses its connection with the jaw, but maintains contact with the cartilaginous nodule which turns into the incus (anvil). The mandibular articulation of the reptiles is maintained, albeit in a different function and in a different position: This is the first divergence from the situation in the lizard. The second divergence is found in the development of the dentary bone which, being no longer blocked in its posterior extension by the articular bone, rejoins the skull to make contact with the squamosum, thereby establishing the typical mammalian articulation.

In the changes of the mammalian developmental programme, compared with those of the reptiles, we recognize a number of well-known mechanisms: controlled cell degeneration, extension and/or reduction with time in the growth of cartilaginous and bony structures, heterochronies, etc. However, these mechanisms are not sufficient to explain everything.

To answer the second question, we can only turn to rather vague assumptions. Should we follows Simpson and assign the directive force to orthoselection, focussing only on the skull which is of interest to us here? This proposition is difficult to accept as it would imply that each phase of this evolution results from selection, with a new stage being of greater selective value than the preceding one, and with the final result, a mammal, being superior to a reptile in all aspects. This subjective statement is contradicted by observations. During hunting, the barn owl is able to hear rodents moving in the undergrowth with an ear of basically reptilian construction through its columella, just as efficiently as through a chain of small bones. The jaw of the Cretaceous *Tyrannosaurus* worked just as well as that of a carnivorous mammal. Our difficulty in defining the effective role of selection leads us to a view which is too restricted in its applications. In the present case, because the articulo-auditory region of the skull underwent a spectacular evolution, our attention is completely focused on it. We tend to forget the whole organism which is the true target of selection. We can be quite certain that particular components of an organism play a more important role in the achievement of a general adaptation than others. These are the main targets of selection and would therefore be predominantly responsible for evolutionary change. They could also "carry along" with them in their modification of the body to its final shape structures of lower adaptive value such as the mandibulo-auditory region. However, when a feature has achieved a high degree of change and has attained an importance which is critical to the functioning of the animal, could it then become the particular target of selection, leading to its full implementation?

Chapter 11. Is Evolution Continuous or Discontinuous?

Appendix 11.1. The Trilobites of New York State (Fig. 11.2)
In this example, the variation within populations is based on the structure of the eyes. They consists of ocellae distributed over the visual surface in dorso-ventral rows in numbers which vary between the different subspecies. The analysis covers specimens from the central and eastern parts of the USA, an area that was covered by a warm, shallow epicontinental sea in the Michigan Basin during the Devonian. This sea extended eastward over a shallow ridge, spreading north–south over the Appalachians and the state of New York along the margin of a continent situated still farther to the east.

The development took place during the mid-Devonian over about 5 million years. Evolution resulted in a reduction in the number of rows of ocellae from 18 to 15. It has to be stressed that the variations analysed took place within populations which are considered to be chronological subspecies of *Phacops rana*. The entire evolution was restricted within a single species, and no speciation is observed. The evolutionary history starts in the west with *Phacops rana milleri* which possessed 18 rows with 9 ocellae each, whereas in the east *Phacops rana crassituberculata* had 6 ocellae per row. The form with 18 rows persisted in the west, and gave rise locally to a form with 15 rows of ocellae *(Phacops rana paucituberculata)* and a restricted geographical distribution. In the state of New York, to the east, the only known older specimen exhibits 18 rows. In higher strata, however, an increased variability becomes noticeable, as the number of rows decrease from 18 to 17 with the disappearance of the first row. In still higher strata, all individuals in the east possess 17 rows of ocellae *(Phacops rana rana)* and this species then persists in the area. During the latter part of the mid-Devonian, a transgression took place and the first specimens of *Phacops rana rana* began to appear in the western sea where they replaced the forms with 18 rows of ocellae, which apparently then became extinct.

A similar reduction in the rows of ocellae occured at the end of the mid-Devonian. Another population from New York State exhibits a decrease from 18 to 15 rows. The variability here includes a gradation from the rows with the largest number of ocellae, through rows with reduced numbers, to their final complete disappearance. As a result, the subsequent populations possessed only 15 rows of ocellae *(Phacops rana norwoodensis)*. Almost immediately, on a geological time scale at least, this form invaded the western seas during a new transgression, taking the place of the forms with 17 rows, which had disappeared during the preceding, Taghanican regression.

The stability of the species is punctuated by short episodes of morphological change, which N. Eldredge explains by the process of paedomorphosis leading to the preservation of juvenile features of the ancestor (neoteny?).

The example shows that the succession of forms in the western sea may not be explained by gradual evolution *in situ*. The morphological changes took place during fairly rapid episodes after long phases of stability. The new features appeared in "peripheral isolated habitats" in the eastern basin, from which the new forms migrated west during transgressions of the sea.

Appendix 11.2. The Molluscs of Bermuda

S. J. Gould [10] has studied continental molluscs of the species *Poecilozonites bermudensis* from the Bermudas in a Quaternary stratigraphic succession covering some 300 years (figs. 11.3 and 11.4). While the study itself was carried out with commendable detail, the interpretation included in the original paper might lead to confusion, as the variations observed are of purely intraspecific nature and do not support the concept of formation of a new species. Gould's scheme is thus not suitable to illustrate this evolution and we have developed a scheme that facilitates a better understanding of the spatial and temporal history of these snails.

The history of *P. bermudensis zonatus* of the Bermudas shows that this species split during an earlier period into two stocks or subspecies which differed in the colour banding pattern of their shells. They developed alongside each other without mixing. The morphological variations of *P. bermudensis zonatus* that occur with time, are weak and it is impossible to assess the contribution of genetic or ecophenotypic effects to this situation. It is, however, possible to show a relationship between morphology and climate, as the variations in shell thickness and size (Fig. 11.4) correlate with climatic fluctuations. In four cases there is a local differentiation of paedomorphic populations; *falsoti*, *sieglindae,* and *bermudensis* in the east, and *siegmundi* in the west. These could result from peripheral isolation in environment that were especially depleted in calcium.

Epilogue: What Has Become of the History of the Giraffe and Its Neck?

Appendix 16.1. The Neck of the Present Giraffe

Palaeontological history of the giraffes (Fig. 16.1). The oldest fossils attributed to the genus *Giraffa* date from the end of the upper Miocene in East Africa, some 10 million years ago. They are assigned to the species *Giraffa jumae,* which was larger than the largest present giraffes *(G. camelopardalis).* In sediments of the late Pliocene, about 2 million years ago, we find remnants which can be assigned to the large form *G. jumae* in eastern and southern Africa, and to the smaller forms *G. gracilis* and *G. pygmea* (East Africa), and eventually to the first remains of the present giraffe in Chad. The legs and neck of *G. gracilis* were as long as those of the present giraffe, but the rest of the body was of a lighter construction with finer proportions. During the lower Pleistocene, these forms still coexisted. The pygmy giraffe was the first to disappear, followed by *G. jumae* and *G. gracilis,* leaving only *G. camelopardalis* to persist to the present day.

It is also of interest to look for the possible ancestor of the giraffe lineage. The giraffe and the okapi of the Congo rainforest are considered as sister groups, the origins of which are still not known. C. S. Churcher suggested that both might have been derived from the African *Samotherium* which, in turn, developed from the more archaic *Palaeotragus*. The former is a giraffide of medium size with simple paired horns. The latter is a small deer-sized giraffide with a shorter neck, possibly without horns, and with barely extended legs.

Taking into account the strong variability and pronounced sexual dimorphism in the present species. C. S. Churcher [2] proposed that the species *G. jumae,* which differs from the present species in the angle between the horns and skull, is undoubtedly a true species on its own. In contrast to this, the assignment of *G. gracilis* and *G. pygmea* to separate species is more debatable.

Considering morphological differences, such as shape and angle between horns and skull, in *G. jumae* and *G. camelopardalis,* as well as the stratigraphic position of the respective fossils, the following evolutionary history may be envisaged:

The genus "*Giraffa*" appeared during the upper Miocene in East Africa where it was represented by the large *G. jumae* which persisted until the end of the mid-Pleistocene. The genus spread during the upper Pliocene and lower Pleistocene north to Chad as *G. pygmea* and to South Africa as *G. jumae* and *G. gracilis*. The present species *(G. camelopardalis)* developed between 3 and 2 million years ago during the upper Pleistocene along the northern edge of its area of occurrence, possibly through allopatric speciation from a source species. It then spread south during the mid-Pleistocene, coexisting with the survivors of its parental species *(G. jumae – gracilis),* and eventually replacing them during the upper Pleistocene.

This type of evolution, in which a parental species gives rise to a descendant species which eventually replaces it in the same area, is a particularly instructive example of the punctuated equilibrium concept of Eldredge and Gould.

If these variations in size and morphology were only of an intraspecific nature, resulting from pronounced sexual dimorphism, it would be necessary to consider the giraffe as representing a single specific lineage, of which the forms *jumae* and *camelopardalis* are only two successive evolutionary stages. The present giraffes all possess markings of the same type, but the shapes and resulting patterns vary between the different subspecies. J. Murray [4] has shown that the patterns may be explained by a reaction–diffusion model during early embryogenesis.

Subject Index

Printing: Mercedesdruck, Berlin
Binding: Buchbinderei Lüderitz & Bauer, Berlin